Environmental Studies: An Integrated Approach

Environmental Studies: An Integrated Approach

Clayton Leon

R CALLISTO
REFERENCE

www.callistoreference.com

Callisto Reference,
118-35 Queens Blvd., Suite 400,
Forest Hills, NY 11375, USA

Visit us on the World Wide Web at:
www.callistoreference.com

ISBN: 978-1-64116-756-7 (Hardback)

Cataloging-in-Publication Data

Environmental studies : an integrated approach / Clayton Leon.
    p. cm.
Includes bibliographical references and index.
ISBN 978-1-64116-756-7
1. Human ecology--Study and teaching. 2. Environmental sciences. 3. Nature study.
4. Environmental sciences--Study and teaching. I. Leon, Clayton.
GF26 .E58 2023
304.2--dc23

Table of Contents

Permissions

Index

Preface

Environmental studies is a multidisciplinary academic field that investigates human interaction with the environment. It combines principles from economics, social sciences, physical sciences and humanities for addressing complicated modern environmental concerns. Enviornmental studies is a comprehensive field of study that encompasses both the built environment and the natural environment along with their interactions. It studies the basic principles of environmental science and ecology. It also includes the study of related subjects such as sociology and social justice, ethics, philosophy, natural resource management, geography, education, anthropology, politics, policy, urban planning, law, pollution control and economics. This book explores all the important aspects of environmental studies in the present day scenario. It aims to serve as a resource guide for students and experts alike and contribute to the growth of the discipline.

This book unites the global concepts and researches in an organized manner for a comprehensive understanding of the subject. It is a ripe text for all researchers, students, scientists or anyone else who is interested in acquiring a better knowledge of this dynamic field.

I extend my sincere thanks to the contributors for such eloquent research chapters. Finally, I thank my family for being a source of support and help.

Clayton Leon

Chapter 1

Multidisciplinary Nature of Environmental Studies

1.1   Definition

Environment is defined as the surroundings or conditions in which a person, animal or plant lives or operates.

Some important definitions of environment given by well-known environmentalist are given below:

Boring:

An environment consists of the sum of the stimulation which receives from conception until death. It can be concluded from the above definition that, "Environment comprises of various types of forces such as physical, political, intellectual and economic, social, cultural, moral and emotional".

Douglas and Holland:

The term environment is used to describe all the external forces, influences and conditions, which affect the life, nature, behaviour, growth, development and maturity of living organisms.

Environmental Study

Environmental study deals with the analysis of the processes in air, water, soil, land and organisms which lead to pollute or degrade the environment.

It helps us for establishing the standard for safe, clean and healthy natural ecosystem. It also deals with important issues like safe and clean drinking water, fertility of land, hygienic living conditions and clean and fresh air, healthy food and development.

The environmental studies include not only the study of physical and biological characters of the atmosphere but also the social and cultural factors and the impact of man on environment.

Environmental property is the ability to keep up the rates of natural resources harvest, pollution creation and non-renewable resource depletion that may be continued indefinitely.

Multidisciplinary Nature of Environmental Studies

The science of environmental studies is a multidisciplinary science as it depends on various disciplines like medical science, physics, chemistry etc. It is the science of physical phenomena in the environment. It is a multidisciplinary field that draws upon not only its core scientific areas, but also applies knowledge from other non-scientific studies such as law, economic and social science.

Chemistry:

To understand the molecular interactions in the system.

Physics:

- To construct the mathematical models of environment.

- To understand the flux of material and energy interaction.

Atmospheric Science:

- It comprises of greenhouse gas phenomena, meteorological studies, airborne contaminants, sound propagation phenomena related to noise pollution and even light pollution.

- To examine the phenomenology of the Earth's gaseous outer layer with emphasis on interrelation to other systems.

Biology:

To describe the effects within the plant and animal kingdom and their diversity.

Environmental Chemistry:

- Principal area of study includes soil contamination and water pollution.

- To study the chemical alterations in the environment.

- Analysis involves multiphase transport of chemicals, chemical degradation in the environment and chemical effects upon biota.

Ecology:

- These studies could also include predator interactions, endangered species, impact analysis of proposed land development upon species viability or effects upon populations by environmental contaminants.

- To analyze the dynamics among an interrelated set of population and some aspects of its environment.

Mathematics and Computer Science:

It will help in environmental modeling and analysis of environment related data.

Geo-science:

- In some classification systems, it can also embrace hydrology including oceanography.

- It includes environmental soil science, environmental geology, volcanic phenomena and evolution of earth's crust.

Law:

It helps in framing of environment related acts, laws, rules and their monitoring.

Economics:

It deals with the economic aspects of various components of environment.

Social Science:

It helps in dealing with the population and health related issues.

1.1.1 Scope and Importance

Scope of Environment

The environment consists of four segments. They are described below:

Atmosphere:

The atmosphere implies the protective blanket of gases, surrounding the earth.

- It sustains life of Earth.
- It saves from the hostile environment of outer space.
- It absorbs most of the cosmic rays from the outer space and a major portion of the electromagnetic radiation from the sun.
- It transmits only near infrared radiation, radio waves, ultraviolet and visible waves. While filtering out tissue damaging ultraviolet waves below 300mm.

The atmosphere is composed of nitrogen and oxygen. Besides, argon, carbon dioxide and trace gases.

Hydrosphere:

The Hydrosphere comprises of all types of water resources such as glaciers, ground water, lakes, oceans, polar icecaps, reservoir, rivers, seas and streams.

- 97% of the earth's water supply is ocean.

- About 2% of the water resources are locked in either polar icecaps or glaciers.

- Only about 1% is available as fresh surface water such as ground water, lakes, rivers, streams and fit to be used for human consumption and other uses.

Lithosphere:

Lithosphere is an outer mantle of the solid earth. It comprises of minerals occurring in the earth's crusts and the soil. Some examples are air, minerals, organic matter and water.

Biosphere:

Biosphere indicates the realm of living organisms and their interactions with the environment, viz lithosphere, hydrosphere and atmosphere.

Importance of Environment Studies

The environment studies enlighten us, about the importance of conservation and protection of our environment from indiscriminate release of pollution. At present a great number of environmental issues have grown by size and complexity, threatening the survival of mankind on earth.

Environment studies have become significant for the following reasons.

Environment Issues Being of International Importance:

It has been well recognized that environmental issues such as acid rain, biodiversity, global warming, marine pollution and ozone depletion are not merely national issues but these are global issues and hence it must be tackled with international efforts and co-operation.

Problems Cropped in the Wake of Development:

Development, in its wake gave birth to agriculture and housing, industrial growth, transportation systems, urbanization, etc. However, it has become phased out in the developed world. The North, to cleanse their own environment has fact fully managed to move 'dirty' factories on South. When the West developed, it did so perhaps neglecting the environmental impact of its activities. Evidently such a path is neither desirable nor practicable, even if developing countries follows that.

Explosively Increase in Pollution:

World census reflects that one in every seven person in this planet lives in India. Evidently with 16% of the world's population and only 2.4% of its land area, there is a heavy pressure on the natural resources including land. Agricultural experts have recognized soil health problems such as deficiency of micronutrients and organic matter, soil salinity and damage of soil structure.

Need for an Alternative Solution:

It is essential especially for developing countries to find alternative paths to attain alternative goals.

Need to Save Humanity from Extinction:

It is our responsibility to save the humanity from extinction. As a result of development, the environment has been constricted and the biosphere has been depleted.

Need for Wise Planning of Development:

Our survival and sustenance depend. Resources withdraw, processing and use of the products have to be synchronized with the ecological cycles in any plan of development, our actions should be planned ecologically for the sustenance of the environment and development.

1.1.2 Sustainability: Stockholm and Rio Summit

The Stockholm and Rio Declarations are the outputs of the first and second global environmental conferences, namely the United Nations Conference on the Human Environment in Stockholm, June 5-16, 1972 and the United Nations Conference on Environment and Development in Rio de Janeiro, June 3-14, 1992. Other policy or legal instruments that emerged conceptually as well as politically from these conferences are the Action Plan for the Human Environment at Stockholm and Agenda 21 at Rio. Adopted twenty years apart, they undeniably represent major milestones in the evolution of international environmental law, bracketing the "modern era" of international environmental law.

Stockholm first represented taking the stock of global human impact on environment, an attempt at forging a basic common outlook on how to address the challenge of preserving and enhancing the human environment. As a result, the Stockholm Declaration supports mostly broad environmental policy goals and objectives rather than detailed normative positions. However, following Stockholm, global awareness on environmental issues increased dramatically, as a part of international environmental law. At the same time, the focus of international environmental activism progressively expanded beyond the trans-boundary and global common issues to media specific and cross sectoral regulation and synthesizing of economic and development considerations in environmental decision making. Therefore, by the time of the Rio Conference, the task for an international community became systematizing and restating the existing normative expectations regarding the environment, as well as of boldly positioning the legal and political underpinnings of sustainable development. In this vein, UNCED was expected to craft an "Earth Charter", a solemn declaration on obligations and legal rights bearing on an environment and development, in the mold of the United Nations General Assembly's 1982 World Charter for Nature. The compromise text that emerged at Rio was not the lofty document an originally envisaged, the Rio Declaration, which reaffirms and builds upon the Stockholm Declaration, has nevertheless proved to be a major environmental legal landmark.

General observations:

The Rio Declaration features a preamble and 27 principles whereas; the Stockholm Declaration consists of a preamble featuring seven introductory proclamations. As diplomatic conference declarations, both the instruments are not formally binding. However, both declarations include the provisions at the time of their adoption were either expected to shape future normative expectations or understood to reflect the customary international law. Moreover, the Rio Declaration, by expressly reaffirming and building upon the Stockholm Declaration, reinforces the normative significance of those concepts common to both the instruments.

The prevention of environmental harm:

Probably the most significant provision that is common to both the declarations relates to the prevention of an environmental harm. In the identical language, the second part of both Rio Principle 2 and Stockholm Principle 21 establishes a state's responsibility to ensure that activities within its control which do not cause damage to the environment of other states or to areas beyond national jurisdiction or control.

The right to development in an environmental context:

Both at Stockholm and at Rio, characterization of the relationship between environment and development was one of the most sensitive challenges facing the respective conference. An initial ecology oriented drafts circulated by the western industrialized countries failed to get traction as developing countries successfully reinserted a developmental perspective in the final versions of the two declarations.

Environmental liability and compensation:

Both the Stockholm and the Rio Declarations calls for the further development of the law bearing on compensation and environmental liability whereas, Stockholm Principle 22 refers to an international law only and the corresponding Rio Principle 13 refers to both national and international law. Notwithstanding these clear mandates, states have tended to shy away from addressing the matter head on or comprehensively, preferring instead to establish the so called private law regimes which focus on private actors liability, while mostly exclude the consideration of states accountability.

1.2   Global Environmental Challenges: Global Warming and Climate Change

The increased input of CO_2 and other greenhouse gases into the atmosphere as a result of human activities will enhance the earth's natural greenhouse effect on raising the average global temperature of the atmosphere near the earth's surface. This enhanced greenhouse effect is called global warming.

Effects of Global Warming

Effect of sea level:

As a result of glacial melting and thermal expansion of the ocean, 20 cm rise in sea level is expected by 2030.

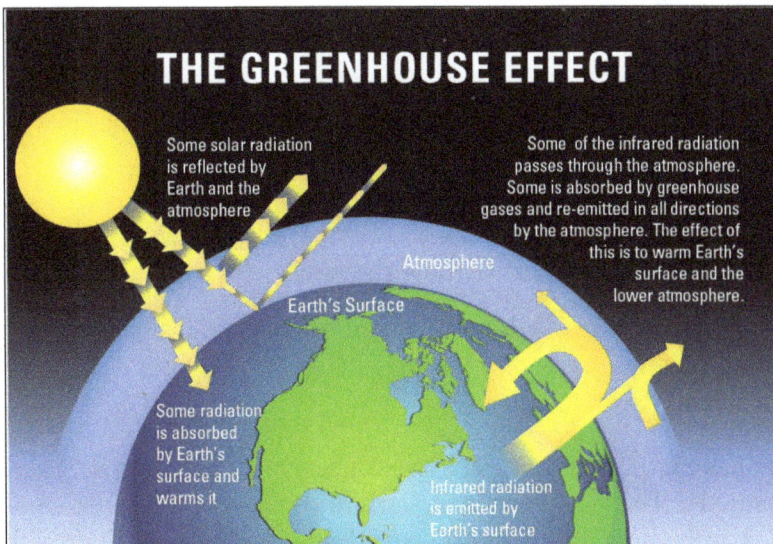

Global warming.

Effects on agriculture and forestry:

High CO_2 levels in the atmosphere have long term negative effects on food production and forest growth. More grain belts would become less productive.

Effects on water resources:

Global rainfall patterns will change and the water management strategies of different region will need to adapt to these changes. Drought and flood will become more common while rising temperature will increase domestic water demand.

Effects of terrestrial ecosystem:

Many plants and animal species will have adapting problems. This will influence the mix of species at different locations. Many will be at risk of extinction whereas, more tolerant varieties will thrive.

Effect on human health:

As the earth becomes warmer, floods and drought become more frequent. There would be increase in water borne diseases, infectious diseases carried by mosquitoes and other disease vectors. The climate change might cause some ecosystem to exceed critical threshold and results in irreversible decline.

Measures to Reduce Global Warming

- Reducing the use of fossil fuels to avoid CO_2 emission.
- Implement energy conservation measures.
- Proper utilization of renewable resources like wind, solar.
- Plant more trees.
- Shift from coal, to natural gas.
- Adopt sustainable agriculture.
- Stabilize population growth.
- Efficiently remove CO_2 from smoke stacks.
- Remove atmospheric CO_2 by utilizing photosynthetic algae.

Climate Change

The activities of the transport industry release several million tons of gases each year into the atmosphere. These include lead, carbon dioxide, methane, carbon monoxide, nitrogen oxides, nitrous oxide, chlorofluorocarbons, perfluorocarbons, silicon tetrafluoride (SF_6), benzene and volatile components, heavy metals and particulate matters. Some of these gases, particularly nitrous oxide, also participate in depleting the stratospheric ozone layer which naturally screens the earth's surface from UV radiation. It is relevant to underline that the climate change also has a significant impact on transportation systems, particularly infrastructure.

Impact of Climate Change

Warming air temperature can raise stream and lake temperatures. It can harm aquatic organisms that live in cold water habitats, such as trout. Warmer water can increase the range of non-native fish species, permitting them to move into previously cold water streams.

The population of native fish species often decreases as non-native fish prey on and outcompete them for food. Impacts of climate change on water availability and water quality will affect many sectors, including energy production, infrastructure, human health, agriculture and ecosystems.

Some regions of the United States, particularly the Northwest, use water to produce energy through hydropower. If a climate change result in a lower stream flows in areas where hydropower is generated, it will reduce the amount of energy that can be produced.

Changes in the timing of stream flow can also have an impact on the ability to produce hydroelectricity. Lower water flows would also reduce the amount of water available to cool fossil fuel and nuclear power plants.

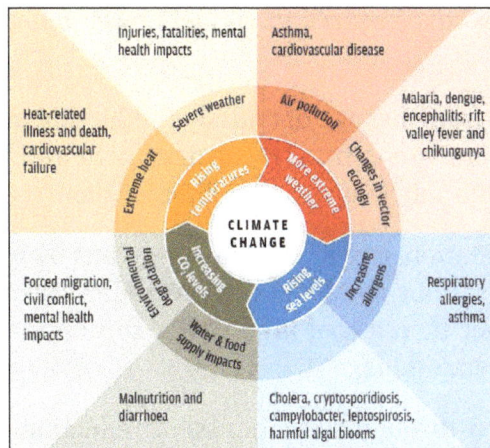

Impact of climate change on human health.

Climate change impacts on water supply and quality will also affect tourism and recreation. Quality of lakes, streams, coastal beaches and other water bodies which are used for swimming, fishing and other recreational activities can be affected by changes in precipitation, increases in temperature and sea level rise. In addition, a winter sport activity that depends on the production of the snow and ice could be limited in the future as its temperature increases.

Agriculture and livestock depends on water. Heavy rainfall and the flooding can damage crops and increase soil erosion and delay planting. Areas that experience more frequent droughts will have less water available for crops and livestock. Aquatic species that live in only cold water environments, such as salmon, is affected by rising water temperatures.

Changing water temperatures would also affect the geographic range of fish species. Changes in the availability and quality of water are major concerns for other countries where water resources are already stressed. Planners in many sectors will confront the challenge of the changing water supply. They are likely to adopt variety of adaptation practices designed to better conserve our water supplies, improve water recycling and develop alternative strategies for water management.

1.2.1 Carbon Credits

Carbon credit is a financial instrument that represents a ton of carbon dioxide or equivalent gases of carbon dioxide removed or reduced from the atmosphere with the help of emission reduction project, which can be used by industry Governments or private institutions to offset damaging carbon emissions that they are generating.

Carbon credits are associated with removing existing CO_2 or CO_2e emissions from the atmosphere in the case of carbon sequestration from forests and planting of trees or the reduction of future CO_2 or CO_2e emissions from energy efficiency projects and renewable energy resources that displace fossil fuel power generation production or industrial processes.

Carbon credits originate from a range of emission reduction activities associated with the removal of existing emissions from the atmosphere and the reduction of future emissions. These are commonly known as "methodologies".

Reforestation and afforestation activities are a key means by which existing emissions can be removed from the atmosphere and carbon credits created during the construction of a wind farm rather than a coal fired power station may create carbon credits through reducing future emissions.

Carbon credits originated through these emission reduction activities can be created under a variety of voluntary and compliance market mechanisms, standards and schemes. Some of these instruments have been established so countries can comply with their mandatory Kyoto targets for voluntary offsetting purposes.

Some schemes around the world offer more environmental benefits than others. Developing parts of the world produce the most carbon credits, often these locations are essentially considered as environmental hot spots as they lack the appropriate regulations, laws and funding that usually exist in developed regions. Due to these reasons they have the most room for improvement and therefore offer the most environmental benefits.

Consistent with all forms of climate change mitigation, in the form of internal abatement, project based reductions, the improved efficiency within businesses or emissions trading schemes, the aim is to achieve reductions based on the lowest possible cost and in line with this theory the Developing World is seen as a low hanging fruit. Projects within these locations and schemes should be additional, which means that the people behind a project needs to demonstrate that the emission reductions would not have occurred without the combined incentives that the carbon credits provided. Due to financial, political or other barriers, the project must prove that it goes beyond a "business as usual" scenario and that the greenhouse gas emissions are lower with the project than without it.

1.2.2 Acid Rains

Acid rain is one of the most dangerous and widespread form of pollution.

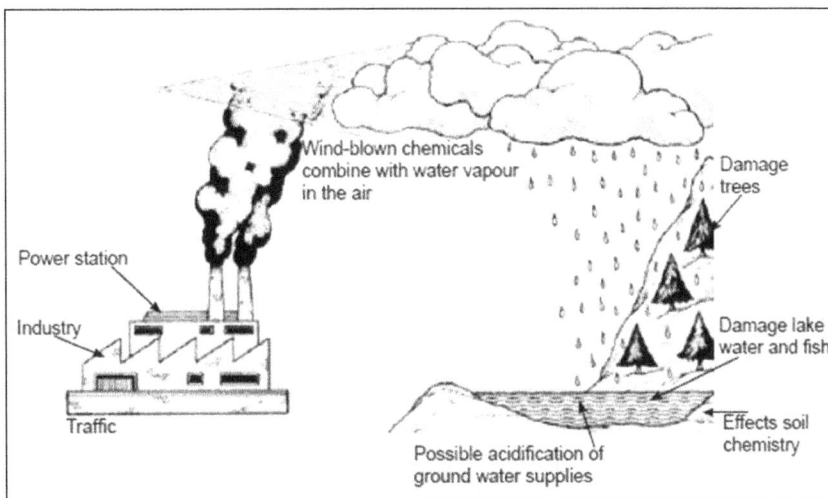

Acid rain cycle.

Sometimes called "the unseen plague", acid rain can go undetected in an area for years. Technically, acid rain is a rain that has larger amount of acid in it than normal.

Acid rain is caused by gases and smoke that are released by factories and cars that run on fossil fuels. When these fuels are burned to produce energy, the sulphur that is present in fuel combines with oxygen and becomes sulphur dioxide. Some of the nitrogen in the air turns into nitrogen oxide. This pollutant goes into the atmosphere and becomes acid.

Acid rain is an extremely destructive form of pollution and the environment suffers from its effects. Forests, trees, lakes, animals and plants suffer from acid rain. Humans will become seriously ill and can even die from the effects of acid rain. One of the major problems that is caused by acid rain is respiratory problems in human being. Many find it difficult to breathe, especially people who have asthma. Asthma, along with dry coughs, headaches and throat irritations can also be caused by the sulphur dioxides and nitrogen oxides from acid rain.

Acid rain can be absorbed by both animals and plants. When the humans eat these plants or animals, toxins in their meals can affect them. Kidney problems, brain damage and Alzheimer's disease has been linked to people eating the "toxic" animals and plants.

Trees are an extremely important natural resource. They regulate local climate, provide timber and forests are homes to wildlife. Acid rain can make trees lose their leaves or needles. The needles and leaves of the trees turn brown and fall off.

Effect of acid rain on forest.

Trees can also suffer from the stunted growth and have damaged bark and leaves, which makes them very much vulnerable to weather, disease and insects. All this happens partly because of direct contact between the trees and acid rain, but it also happens when the trees absorbs soil that has come in contact with acid rain. The soil poisons the trees with toxic substances that acid rain has deposited into it.

Lakes are also damaged by acid rain. A lake polluted by acid rain will support only the hardiest species. Fishes dies off and that removes the main source of food for the birds. Also, birds can die from eating the "toxic" fish and insects and vice versa.

Acid rain can even kill the fish before they are born. Acid rain hits the lakes mostly in the springtime, when fish lay their eggs. The eggs come in contact with the acid and entire generation can be killed. Fish generally die only when the acid level of a lake is high. When the acid level is lower, they become sick, suffer stunted growth or lose their ability to reproduce.

Castle stonework in the UK damaged by acid rain.

Artwork and architecture can be destroyed by acid rain. Acid particles can land on buildings, causing corrosion. When sulphur pollutants fall on the surfaces of buildings, they react with the minerals in the stone to form a powdery substance that can be washed away by rain. This powdery substance is known as gypsum. Acid rain can damage airplanes, buildings, cars, railroad lines, stained glass, steel bridges and underground pipes.

The below picture shows how acid rain has eroded the stonework of a castle in Lincolnshire, England and has left a figurehead barely visible.

Stonework of a castle in Lincolnshire, England damaged by acid rain.

Currently, both airplane industry and railway industry have to spend a lot of money to repair the corrosive damage done by acid rain. Also, bridges have collapsed in the past due to acid rain.

1.2.3 Ozone Layer Depletion

An ozone molecule is composed of three atoms of oxygen. Ozone in the upper atmosphere is known as "ozone layer", it protects life on Earth by absorbing most of the ultraviolet radiation emitted by the sun. Exposure to too much UV radiation is linked to cataracts, skin cancer and depression of the immune system and may reduce the productivity of certain crops. Accordingly, stratospheric ozone is termed as "good ozone." In contrast, human industry creates "ozone pollution" at the ground level. This "bad ozone" is a prime component of smog.

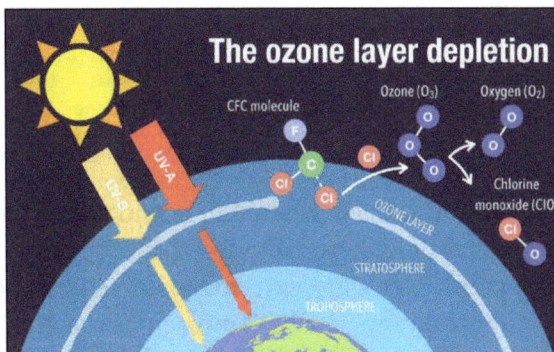

Ozone layer depletion.

The ozone layer is reduced when man made CFC molecules reach the stratosphere and are broken by short wave energy from the sun.

Free chlorine atoms then break apart molecules of ozone, creating a hole in the ozone layer. The hole in the ozone layer over the Antarctic in 1998 was the largest observed since the annual holes first appeared in the late 1970s.

The CFCs were once used in as foam blowing agents and in aerosol sprays. Their manufacture is now banned by an international treaty, the Montreal Protocol, signed by 160 nations. As CFCs have a long atmospheric lifetime, those manufactured in the year 1970s continue to damage the ozone layer today.

The good news is that scientists predict that the ozone layer will return to its earlier stable size by the middle of the 21st century assuming that nations continue to comply with the treaty.

When the ozone hole was first detected, there was an emotional debate in which many U.S. industries fiercely resisted a ban of CFCs. It took a few years for scientists to show conclusively that a human activity was causing the damage. It did not take too long for scientists to invent other chemicals that could replace CFCs for commercial and industrial purposes, but would not harm the ozone layer. CFCs used as propellants were first banned in the United States in 1978.

Process of Ozone Depletion

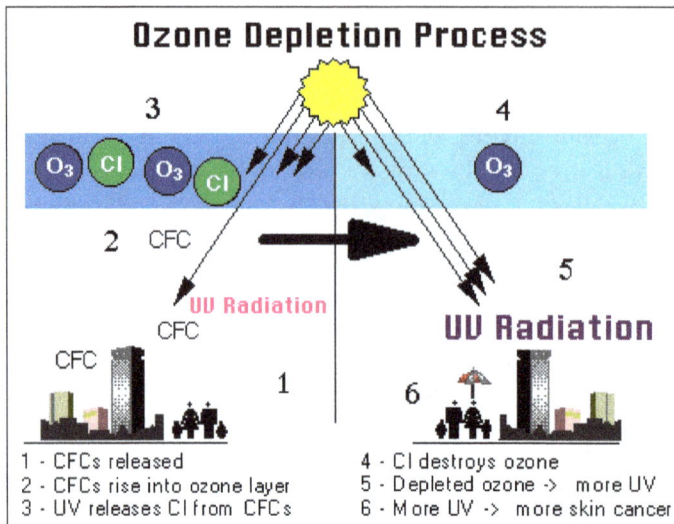

The process of ozone depletion begins when CFCs and other Ozone Depleting Substances are emitted into the atmosphere. Winds efficiently mix with the troposphere and evenly distribute the gases. CFCs are extremely stable and they do not dissolve in the rain. After a period of several years, an ODS molecules reach the stratosphere which starts at about 10 kilometers above the Earth's surface.

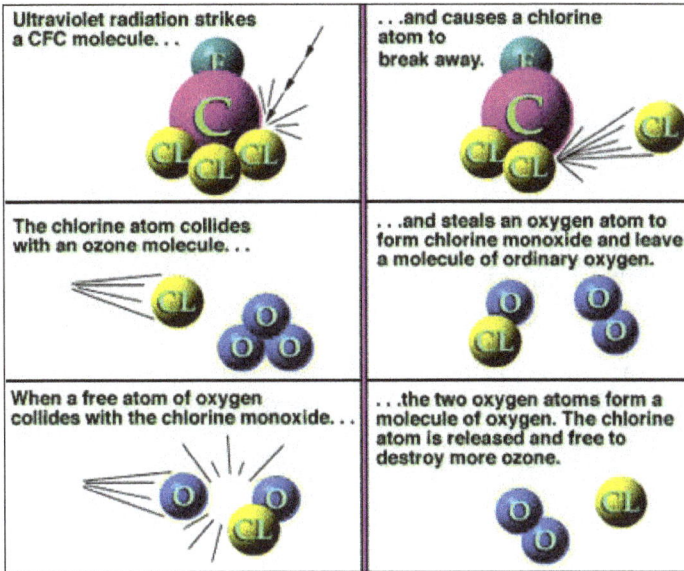

Ultraviolet radiation strikes a CFC molecule...	...and causes a chlorine atom to break away.
The chlorine atom collides with an ozone molecule...	...and steals an oxygen atom to form chlorine monoxide and leave a molecule of ordinary oxygen.
When a free atom of oxygen collides with the chlorine monoxide...	...the two oxygen atoms form a molecule of oxygen. The chlorine atom is released and free to destroy more ozone.

Strong UV light breaks the ODS molecule. CFCs, HCFCs, carbon tetrachloride, methyl chloroform and other gases release chlorine atoms and halons and methyl bromide release bromine atoms. It is these atoms that destroy ozone, not the intact ODS molecule. It is estimated that about one chlorine atom can destroy about 1,00,000 ozone molecules before it is removed from the stratosphere.

Ozone is constantly produced and destroyed in a natural cycle, as shown in the above figure. However, the overall amount of ozone is essentially stable. This balance can be seen of as a stream's depth at a specific location. Although individual water molecules are moving past the observer, the total depth remains constant whereas, ozone production and destruction are balanced at ozone levels remains stable. This was the situation until the past several decades that have upset the balance. In this effect, they have added a siphon downstream, removing the ozone faster than the natural ozone creation reactions can keep up. Therefore, ozone levels fall.

Since ozone filters out harmful Ultraviolet B radiation, less ozone means higher UVB levels at the surface, larger increase in incoming UVB. UVB has been linked to cataracts, skin cancer, and damage to materials like plastics and harm to crops and certain marine organisms. Although some UVB reaches the surface even without ozone depletion, its harmful effects will increase as a result of this problem.

1.2.4 Population Growth and Explosion

Population Growth and Environmental Issues

The rapid growth of global population for the past 100 years results from the difference between the rate of birth and death. In 1980, global population was about 1 billion people. It took about 130 years to reach 2 billion. But the population reached to 4 billion within 45 years. Now we have already crossed 6 billion and may have to reach about 10 billion by the year 2050 as per the World Bank calculations.

Causes of Rapid Population Growth

- The rapid population growth is due to decrease in death rate and increase in birth rate.

- The availability of antibiotics, immunization, increased clean water, food production and air decreases the famine related deaths and infant mortality.

- In agricultural based countries, children are required to help parents in the fields that is why population increases in the developing countries.

Characteristics of Population Growth

Exponential growth:

Now population growth occurs exponentially like 10, 10^2, 10^3, 10^4, etc., which shows the dramatic increase in global population in the past 160 years.

Doubling time:

It is the time required for a population to double its size at a constant annual rate. If a nation has 2% annual growth, its population will double in next 35 years.

Infant mortality rate:

It is the percentage of infants died out of those born in the year. Even though this rate has decreased in the last 50 years, the pattern differs widely in developing countries.

Total fertility rates (TFR):

It is the average number of children delivered by a woman in her lifetime. The TFR value varies from 2 in developed countries to 4.7 developing countries.

Replacement level:

Two parents bearing two children will then be replaced by their off spring. Due to infant mortality their replacement level is changed. But, due to high infant mortality replacement level is generally high in developing countries.

Male - Female ratio:

The ratio of girls and boys should be fairly balanced in a society to flourish. But the ratio has been upset in many countries including China and India. In China, the ratio of girls and boys is 100:140.

Demographic transition:

Population growth is generally related to economic development. The death rates and birth rates fall due to improved living condition. This results in low population growth. This phenomenon is referred to as demographic transition.

Problems of Population Growth

- Increasing demands for the food and natural resources.

- Inadequate processing and health services.

- Loss of agricultural lands.

- Unemployment and social political unrest.

- Environmental pollution.

Population Growth Variations

From the dawn of mankind to the turn of the 19th century world population grew to a total of one billion people. During the year 1800s, human numbers increased at increasingly higher rates, reaching a total of about 1.7 billion people by the year 1900. World population has grown even more rapidly during the present century, with the greatest gains occurring in the post World War II period and stands at over three times its size in 1900 (i.e.) some 5.9 billion people today.

Population growth has continued throughout the past three decades in spite of the decline in fertility rates that began in many developing countries in late 1970s and the toll taken by HIV/AIDS pandemic. While the rate of increase is declining, in absolute terms world population growth continues to be substantial. Global population increase is equivalent to adding the population of Egypt, Gaza, Israel and Jordan to the existing world total each year.

According to Census Bureau projections, world population will increase to the level of nearly 8 billion persons by the end of next quarter century and will reach 9.3 billion persons, a number more than half again as large as today's total by 2050.

The future of human population growth has been evaluated and is now largely being decided, in the world's Less Developed Nations. 96% of world population increase now occurs in developing regions of Africa, Asia and Latin America and this percentage will rise over the next quarter century.

90% of the world's births and 77% of its deaths will take place in 1998. Census Bureau's projections indicate that early in the next century, crude death rates will exceed crude birth rates for the world's More Developed Countries and the difference natural increase will be negative.

At this point, international migration will become a critical variable determining whether the total population of today's MDC's increases or decreases. These projection shows negative natural increase offset by net international immigration through 2019 but, if present trends continue, the population of the world's MDC's will slowly begin to decrease from the year 2020 onward.

Variation Among Nations

As the growth rate in world's more affluent nations becomes negative, all of the net annual gain in global population will come from the world's developing countries.

Below is a listing of most populated countries in the world:

- Bangladesh → 147,365,352.
- Brazil → 188,078,227.
- China → 1,313,973,713.
- Germany → 82,422,299.
- India → 1,095,351,995.
- Indonesia → 245,452,739.
- Japan → 127,463,611.
- Mexico → 107,449,525.
- Nigeria → 131,859,731.
- Pakistan → 165,803,560.
- Philippines → 89,468,677.
- Russia → 142,893,540.
- United States → 298,444,215.
- Vietnam → 84,402,966.

Population Explosion

The enormous increase in population, due to low death rate and high birth rate is known as population explosion. The human population is not increasing at a uniform rate in all parts of the world.

- Population explosion leads to environmental degradation.
- It causes over exploitation of natural resources. Therefore there will be a shortage of resources for the future generation.
- Many of the renewable resources like forest, grasslands are also under threat.
- The increase in population will increase disease economic inequity and communal war.

Causes of Population Explosion

- Modern medical facilities reduces death rate and increases birth rate.

- Increase of life expectancy.

- Illiteracy.

Effects of Population Explosion

- It leads to environmental degradation.

- It causes over-exploitation of natural resources.

- Increase in population increases disease and communal war.

- Lack of basic amenities like water, education, health etc.

- Overcrowding leads to development of slums.

- Poverty results in infant mortality, is the tragic indicator of poverty.

- Renewable resources like forests are under threat.

Remedy:

Reducing fertility rate through birth control programs.

Family Welfare Programme:

Family welfare programme was implemented by the Government of India as a voluntary programme. It is an integral part of the overall national policy of growth covering maternity, human health, child care, family welfare and woman's right.

Objectives of Family Welfare Programming:

- Slowing down the population explosion by reducing the fertility.

- Pressure on the environment due to over exploitation of natural resource.

Population Stabilization Ratio:

The ratio is derived by dividing crude birth rate by crude death rate.

Developed Countries:

The stabilization ratio of the developed countries, which is more or less stabilized indicating zero population growth.

Developing Countries:

The stabilization ratio of the developing countries nearing 3 is expected to lower down by 2025. Stabilization in developing countries is possible only through various family welfare programs.

1.2.5 Role of Information Technology in Environment and Human Health

Software for environmental education:

Remote Sensing:

Remote sensing refers to any method, which can be used to gather information about an object without actually coming in contact with it. At present, the term remote sensing is used more commonly to denote identification of Earth's features by detecting the characteristic electromagnetic radiation that is reflected and emitted by the Earth.

Example: The remote sensing image of land can be used to derive information of vegetative cover, water bodies, land use, etc.

Application of Remote Sensing

In agriculture:

In India, the agriculture sector sustains the livelihood of around 70% of the population and contributes to about 35% of the net national product.

In forestry:

Sustainable forest management requires reliable information on the type, density and extent of forest cover, wood volume and biomass.

In land cover:

Remote sensing data is covered to map; the spatial resolution plays a vital role on the scale of mapping.

Water Resources

Remote sensing data has been used in many applications related to water resources such as surface water body mapping, wet land and many more.

Database

Database is the collection of interrelated data on various subjects. In computer, the information of the database is arranged in a systematic manner which is easily manageable and can be very quickly retrieved.

Applications

It is used in:

- The Ministry of Environment and Forest.

- National Management Information System (NMIS).

- Environmental Information System (ENVIS).

Geographical Information System

GIS is a technique of superimposing various thematic maps using digital data on a large number of interrelated aspects.

Application of GIS

- Interpretations of polluted zones, degraded lands can be made based on GIS.

- GIS can be used to check unplanned growth.

Satellite Data

- Satellite data helps in providing correct and reliable information about forest cover.

- It also provides information of automation of phenomena like monsoon, ozone layer depletion, smog, etc.

- From one satellite data many new reserves of oil, minerals can be discovered.

- World Wide Web: More current data is available on World Wide Web.

Role of IT in Environment and Human Health

Application of IT in Human Health

- It helps the doctor to monitor the health of the people effectively.

- The information regarding the outbreak of epidemic diseases can be conveyed easily.

- Expert doctors can be consulted through online to provide better treatment and services to the patient.

- With a central control system the hospital can run effectively.

- Drugs and its replacement can be administered effectively.

Environment and Human Health

Human health and environment are two inseparable entities. Generally a physically fit person, not suffering from any disease, is called a healthy person. The factors determining human health are as follows:

- Nutritional factors.

- Biological factors.

- Chemical factors.

- Psychological factors.

Physical hazardous and their health effects:

Physical hazards	Health effects
Radioactive radiations	Attacks the cells in the body and the function of glands and organs.
UV radiations	Skin cancer
Global warming	Temperature increase cause famine, mortality.
Chlorofluorocarbons	Damage O_3 layer, UV rays cause skin cancer.
Noise	Painful and irreparable damage to human ear.

Chemical hazards and their health effects:

Chemical hazards	Health effect
Combustion of fossil fuels: Liberates SO_2, NO_2, O_2 and industrial effluents	Asthma, bronchitis and other lung disease kills cells and causes cancer.
Pesticides like DDT, chlorinated pesticides and heavy metals like Hg, Pd, N etc.	Affects the food chain and contaminates water.

Biological hazards and their health effect:

Biological hazards	Health effects
Bacteria, viruses and parasites	Diarrhoea, malaria, parasites, worms, anemia, cholera etc.

Preventive Measures:

- Before cooking, wash the vegetables and fruits with clean water.

- Drink chemically treated and filtered water.

- Do physical exercise to have proper blood circulation in the body.

- Maintaining the skin, teeth and hair.

- Try to avoid plastic containers and aluminium vessels.

- We should always wash our hands before sitting for food.

- We should cut short and clean our nails.

1.3   Ecosystems: Concept of an Ecosystem

A group of organisms interacting among them and with the environment is known as ecosystem. Thus, an ecosystem is a community of different species interacting with one another and with their non-living environment exchanging energy and matter.

Example: Animals cannot synthesize their food directly but depends on the plants either directly or indirectly. Few examples of ecosystem are pond, forest, estuary and grassland.

Primary Types of Ecosystems

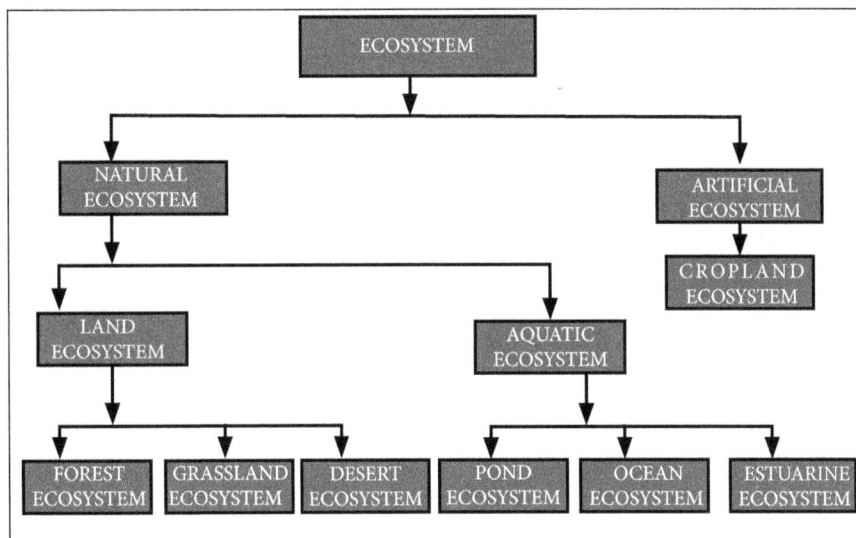

Natural Ecosystems

Natural ecosystems may be terrestrial or aquatic. A natural ecosystem is a biological environment that is found in nature rather than created or altered by man.

Artificial Ecosystems

Humans have modified some ecosystems for their own benefit. These are the artificial ecosystems. They can be terrestrial or aquatic.

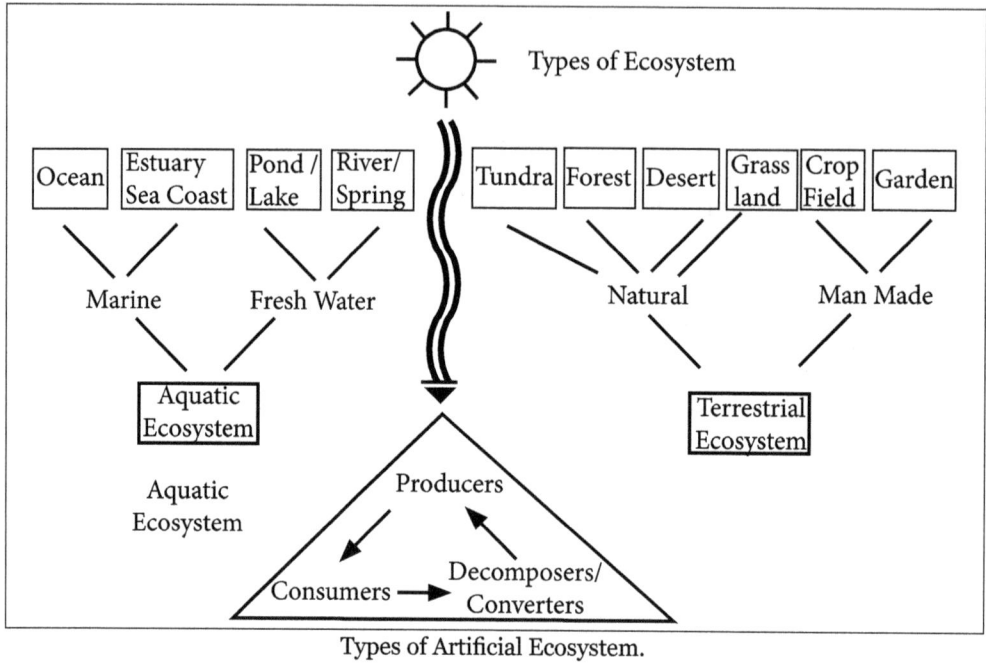

Types of Artificial Ecosystem.

Types of Natural Ecosystems

There are two main types of natural ecosystems. They are:

- Aquatic natural ecosystem.

- Terrestrial natural ecosystem.

In aquatic ecosystems, organisms interact with water. In terrestrial ecosystems, organisms interact with land.

Aquatic Ecosystems

Aquatic ecosystems in general covers up to 71% of the earth's surface. As a type, aquatic ecosystems can be classified again into three varieties, defined by the kind of water with which the organisms interact.

Freshwater: This type of ecosystem includes rivers, lakes, streams, ponds and wetlands and makes up smallest percent of the earth's aquatic ecosystem.

Transitional Communities: These are places where freshwater and salt water comes together, including estuaries and wetlands.

Marine: More than 70% of the earth is covered by salt water or marine ecosystems. These include shorelines, coral reefs and open ocean.

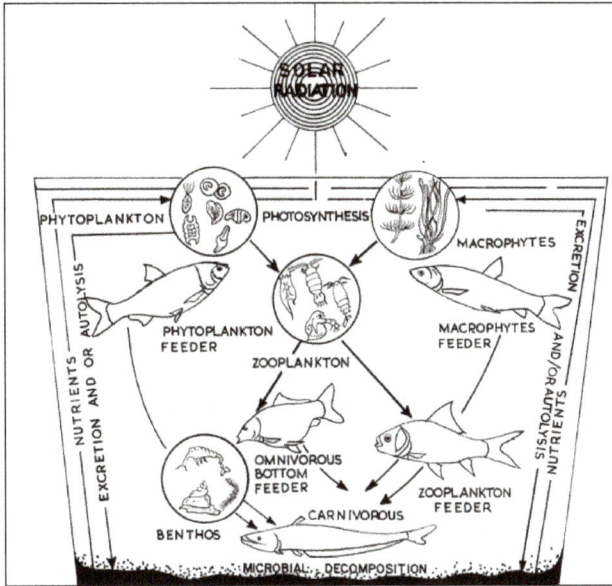

Pond as an ecosystem.

Terrestrial Ecosystems

Terrestrial ecosystems are classified by the type of land or terrestrial area. Mountains, forests, deserts and the grasslands are types of terrestrial ecosystems.

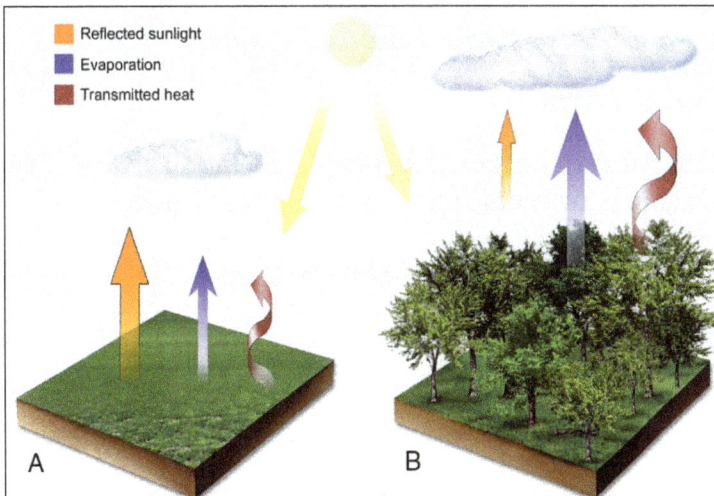

Terrestrial ecosystems.

Forest: These ecosystem features dense tree population and include tropical rain forests.

Desert: Deserts receive less than 25 cm of rainfall per year.

Grassland: These ecosystems include tropical savannas, temperate prairies and arctic tundra.

Mountain: Mountain ecosystem includes steep elevation changes between meadows, ravines and peaks.

Balanced Ecosystem

The term ecosystem describes both the living and non-living components of an area that interact with one another. All the components are interdepartmental in some way with each other. An ecosystem may be aquatic or terrestrial.

In an aquatic ecosystem rocks are needed for shelter and plants provide oxygen for fish. An ecosystem is balanced when natural animals and plants and non-living components are in harmony i.e. there is nothing to disturb the balance.

Advantages

- It is difficult to find a perfectly balanced ecosystem but we can make a model at home and observe how the ecosystem functions. We can observe how different species interact with each other.

- We can study the natural cycle of each species.

- Understand the relationship between different species such as producer, predator and prey.

1.3.1 Structure and Function of an Ecosystem

Structure of an Ecosystem

The term structure refers to various components. So the structure of an ecosystem explains the relationship between the biotic and abiotic components.

An Ecosystem has Two Major Components

- Biotic components.

- Abiotic components.

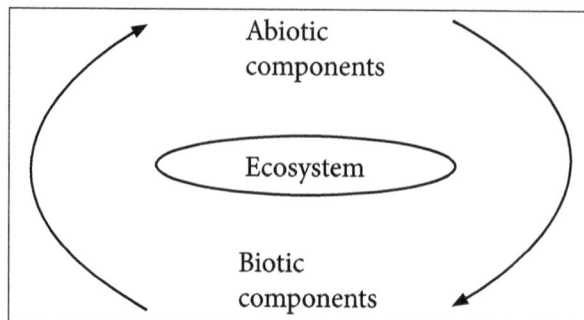

Components of ecosystem.

Biotic Components

The living organisms or the living members in an ecosystem collectively form its community called biotic components or biotic community.

Example: Plants, animals and micro-organisms.

Classification of Biotic Components

The members of the biotic components of an ecosystem are grouped into three groups based on how they get their food:

- Producers (Plants).

- Consumers (Animals).

- Decomposers (Micro-organisms).

Grass	→	Rat	→	Cat	→	Tiger
Producers		Herbivores		Primary Carnivores		Secondary Carnivores

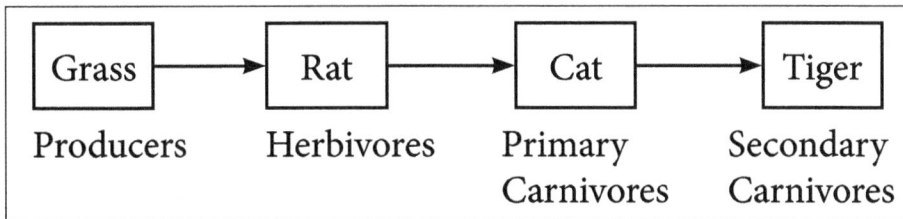

Biotic components of an ecosystem.

Abiotic Components

The non-living components of the ecosystem collectively form a community called abiotic components or abiotic community.

Example: Climate, soil, water and air.

Physical components:

They include the climate, energy, raw materials and living space that the biological community needs. They are useful for the growth and maintenance of its member.

Example: Air, water, soil, sunlight, etc.

Chemical components:

They are the sources of essential nutrients.

Example:
- Organic substances: Protein, carbohydrates, lipids, etc.
- Inorganic substances: All micro and macro nutrients and few other elements.

Functions of an Ecosystem

The function of an ecosystem is to allow flow of energy and cycling of nutrients.

Functions of an ecosystem are of three types. They are as follows:

- Primary function: The primary function of all ecosystems is the manufacture of starch.

- Secondary function: The secondary function of all ecosystems is distributing energy in the form of food to all consumers.

- Tertiary function: All living systems die at a particular stage. These dead systems are decomposed to initiate the third function of ecosystem known as "cycling".

The functioning of an ecosystem can be understood by studying the following terms:

- Energy and material flow.
- Food chains.
- Food webs.
- Food pyramids.

1.3.2 Producers, Consumers and Decomposers

Producers

Producers synthesis their food themselves through the process of photosynthesis. Plants are known as producers, because they produce their own food. They do this by using carbon dioxide from the air, light energy from the sun and water from the soil to produce food in the form of glucose or sugar. This process is called photosynthesis.

Example: All green plants and trees.

Consumers

Consumers are organisms, which cannot prepare their own food and depend directly or indirectly on the producers. Animals are called consumers, because they cannot make their own food, so they need to consume plants or animals. Humans are also omnivores.

Example:

- Plant eating species like insects, rabbit, goat, etc.
- Animals eating species like lion, tiger, etc.

Group of Consumers

There are three groups of consumers. They are:

- Animals that eat only plants.

- Animals that eat only animals.

- Animals that eat both plants and animals.

Classification of Consumers

Consumers are further classified into:

Primary consumers:

Primary consumers are also called herbivores; they directly depend on the plants for their food. So they are known as plant eaters.

Example: Insects, rat and goat.

Secondary consumers:

Secondary consumers are the primary carnivores, they feed on primary consumers. They directly depend on the herbivores for their food.

Example: Frog, cat etc.

Tertiary consumers:

Tertiary consumers are secondary carnivores, they feed on secondary consumers. They directly depend on the primary carnivores for their food.

Example: Tigers, lions, etc.

Decomposers

Decomposers attack the dead bodies of producers and consumer and decompose them into simpler compounds. During decomposition, inorganic nutrients are released. These inorganic nutrients together with other organic substances are then utilized by the producers for the synthesis of their own food.

Example: Micro-organisms like bacteria and fungi.

Bacteria and fungi are decomposers. They eat decaying matter, dead animals and plants and in the process they break them down and decompose them. Then they release nutrients and mineral salts back into the soil which will be used by plants.

1.4   Energy Flow in the Ecosystem

- Solar energy is transformed to chemical energy during the process of photosynthesis.

- Some of the chemical energy is used by plants for their growth and remaining is transferred to other tropic levels.

- Tropic levels are feeding levels or various steps through which the energy passes in an ecosystem.

- The plants use helium and heliotaxes are used by carnivores as their food.

- Finally plants and herbivores are decomposed and then the energy goes to the environment.

- Sun → Plants → Animals → Bacteria.

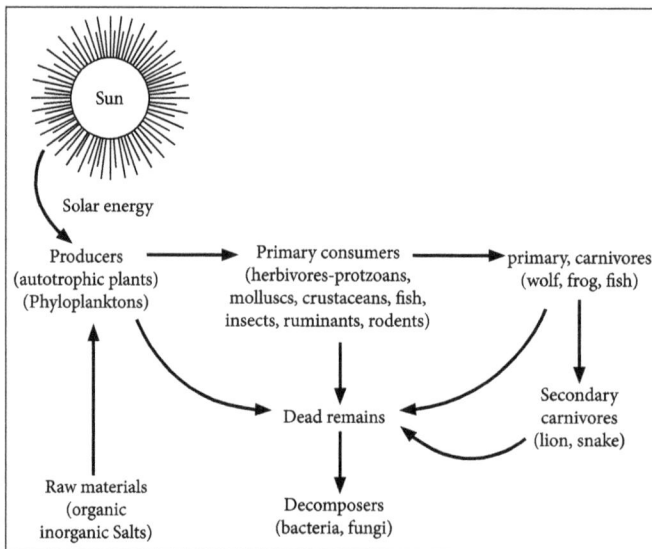

Energy flow at different levels of an ecosystem.

1.4.1 Ecological Succession

The progressive replacement of one community by another life along the development of the stable community in a particular area is called ecological succession.

Stages of Ecological Succession

Pioneer community:

The first group of organism which establishes their community in the area is termed as "Pioneer community."

Seres or seral stage:

The various developmental stages of the community are called as "seres".

Community:

It is the group of plants or animals living in an area.

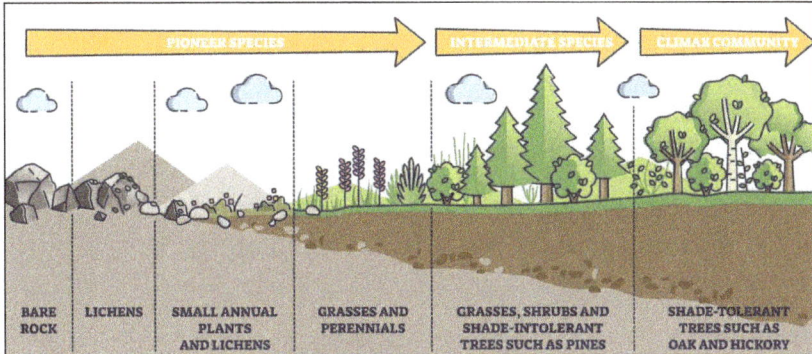

Ecological succession.

Types of Ecological Succession

Ecologists recognize two types of ecological succession based on the conditions present at the beginning of the process such as:

Primary succession:

It involves the gradual establishment of biotic communities on a lifeless ground.

- Hydrarch or Hydrosere: Establishment starts in a watery area like pond or lake.
- Xerarch: Establishment starts in a dry area like deserts and rocks.

Secondary succession:

It involves the establishment of biotic communities in an area where some type of biotic community is already present.

Process of Ecological Succession

The process of ecological succession can be explained in following steps:

Nudation:

It is the development of a bare area without any life form.

Invasion:

A biological invader is a species of animal, plant or microorganism which most usually transported in or intentionally by man and spreads into new territories. Some distance from its home territory. If one looks at long time scales then invasions are probably a frequent phenomenon. However, human activity has increased the frequency of invasions dramatically by disrupting biogeographic barriers and increasing exchanges.

- Migration: Migration of seeds is brought about by water, wind or birds.

- Establishment: The seed then germinate and grow on the land and establishes their pioneer communities.

Competition:

As the number of individual species grows, there arises competition with the same species and between different species for water, space and nutrients.

Reaction:

The living organisms take water, nutrients and grows and modifies the environment is known as reaction. This modification becomes unsuitable for the existing species and also favors some new species, which replace the existing species.

Stabilization:

It leads to stable community, which is in equilibrium with the environment.

1.5   Food Chains, Food Webs and Ecological Pyramids

Food Chains and Food Webs

The most common interactions between species are through feeding relationships. The easiest way to display these relationships is with food chains. Food chains illustrate who eats whom in an ecosystem. In the figure shown given below, the seeds of a producer are eaten by a herbivore which is in turn eaten by a carnivore. Finally, they fall as a prey to a top carnivore.

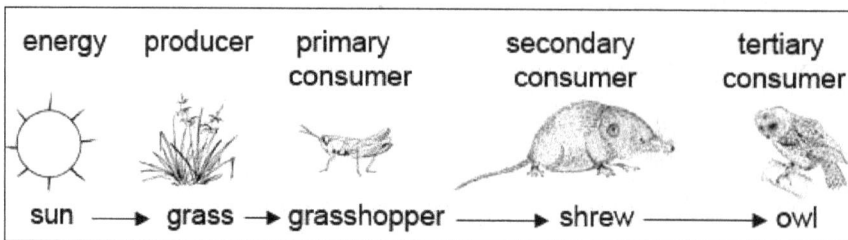

| energy | producer | primary consumer | secondary consumer | tertiary consumer |

sun ⟶ grass → grasshopper ⟶ shrew ⟶ owl

This means that energy is continuously lost from all levels of the food chain.

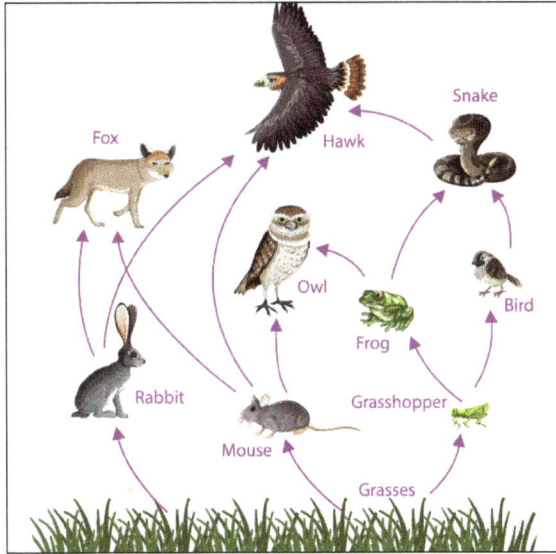

The ecologists refer to the trophic level or feeding level, to describe the position of an organism along a food chain. Producers occupy the lowest or first, trophic level. Herbivores occupy the second trophic level and carnivores occupy the third and fourth trophic level the figure shows given below:

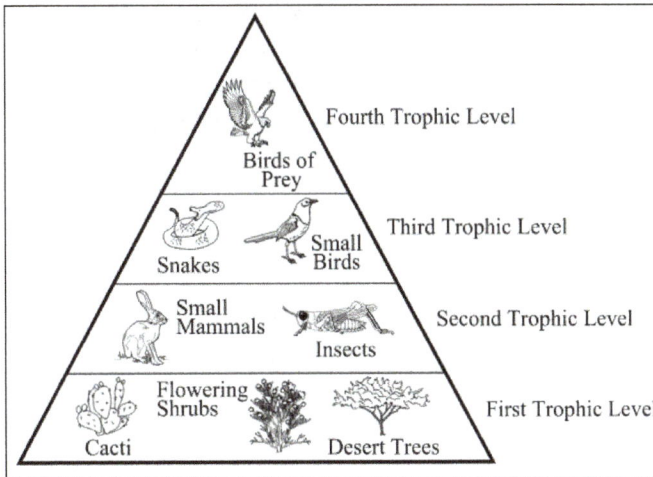

The food chains do not exist in nature. They are used to show simple feeding relationships. Food chains are part of more complex sets of relationships that exist among species. Many herbivores eat pine seeds. Red squirrels eat a wide range of foods and are themselves food for a variety of predators. A more accurate way to illustrate interactions is using a food web. This shows a series of interconnecting food chains

Food webs are highly complex, with consumers feeding on many species if the figure given below. The large number of interactions tends to reduce the vulnerability of any one species to the loss or decline of another species. For this reason, complex food webs are considered as more stable than simple food webs.

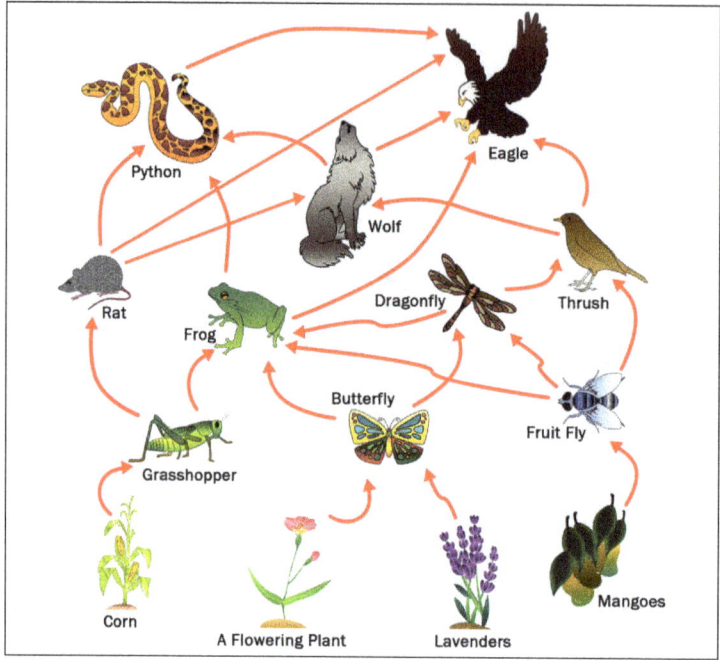

Food web.

The food webs are useful tools to figure out what may happen when a species is removed from or added to an ecosystem. For the example, if a species is removed from a food web, the species it feeds on may increase dramatically in numbers. Conversely, the population of a newly introduced species may disrupt the entire food chain as shown in the figure given below:

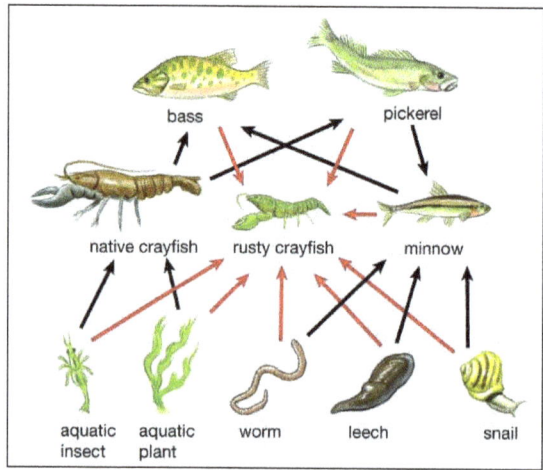

Ecological Pyramids

Graphical representation of structure and function of tropic levels of an ecosystem, starting with producers at the bottom and each successive tropic level forming the apex is known as an ecological pyramid.

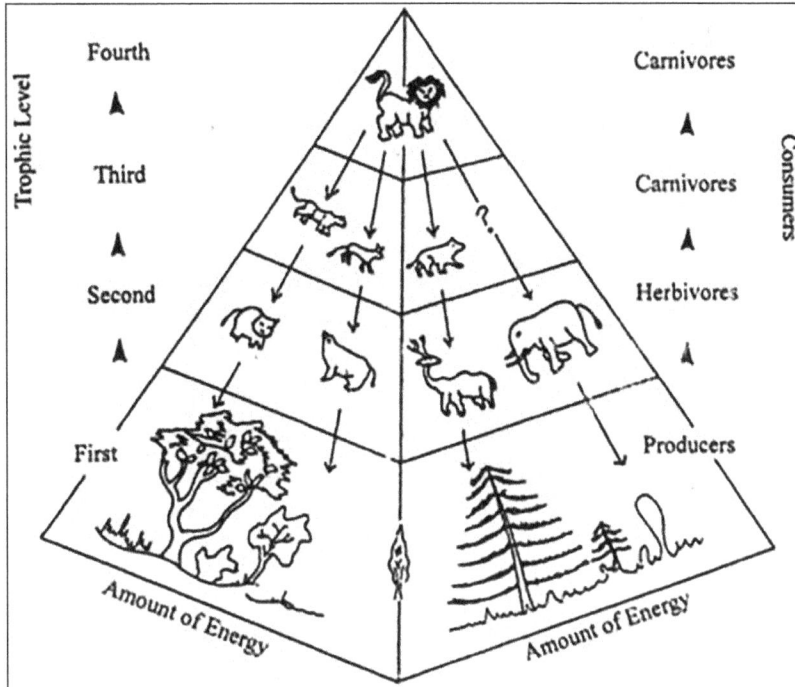

Ecological pyramids.

In the food chain starting from producers to consumers, there is a regular decrease in the properties. Since some energy is lost as heat in each tropic level, it becomes progressively smaller near the top.

Types of Ecological Pyramids

Ecological pyramids are of three types:

Pyramid of Numbers:

It represents the number of individual organisms present in each tropic level.

Example: A grassland ecosystem.

The producers in the grasslands are grasses, which are small in size and large in numbers. So the producers occupy lower tropic levels.

The primary consumers are rats, which occupy the II tropic level. Since the numbers of rats are lower when compared to the grasses, the size of which is lower.

The secondary consumers are snakes, which occupy the III tropic levels. Since the number of snakes are comes when compared to the rats, the size of which is lower.

The tertiary consumers are eagles, which occupy the next tropic level. The number of size of the last tropic level is lowest.

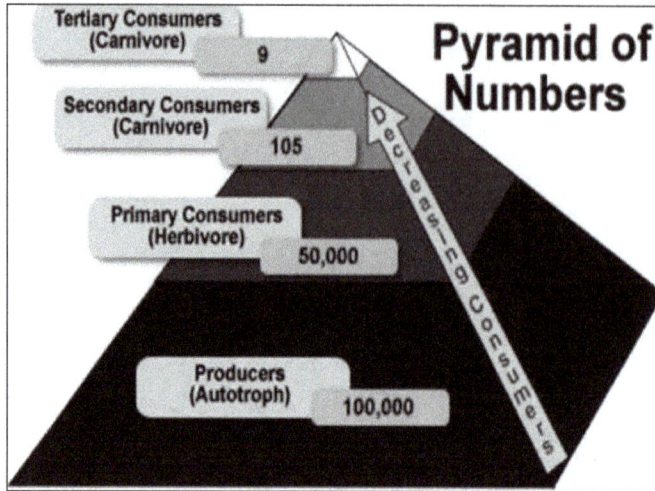

Pyramid of numbers.

Pyramids of Energy:

This represents the amount of energy present in each tropic level. The rate of energy flow and the productivity at each successive tropical level is shown.

At every successive tropic level, there is a heavy loss of energy in the form of heat. Thus at each next higher tropic level only 10% of the energy is transferred. Hence, there is a sharp decrease in energy at each and every successive tropic level as we move from producers to top carnivores.

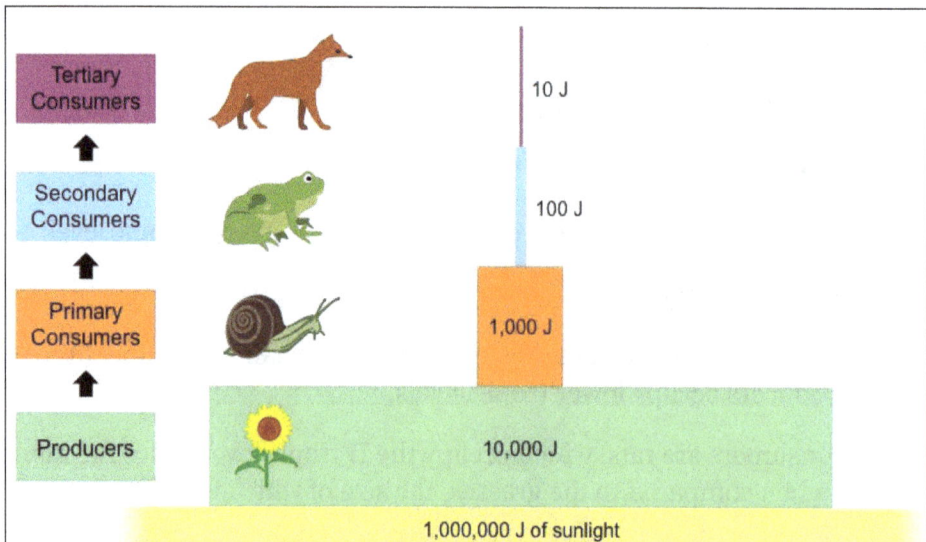

Pyramid of energy.

Pyramids of Biomass:

It represents the total amount of biomass present in each tropic level.

Example: A forest ecosystem.

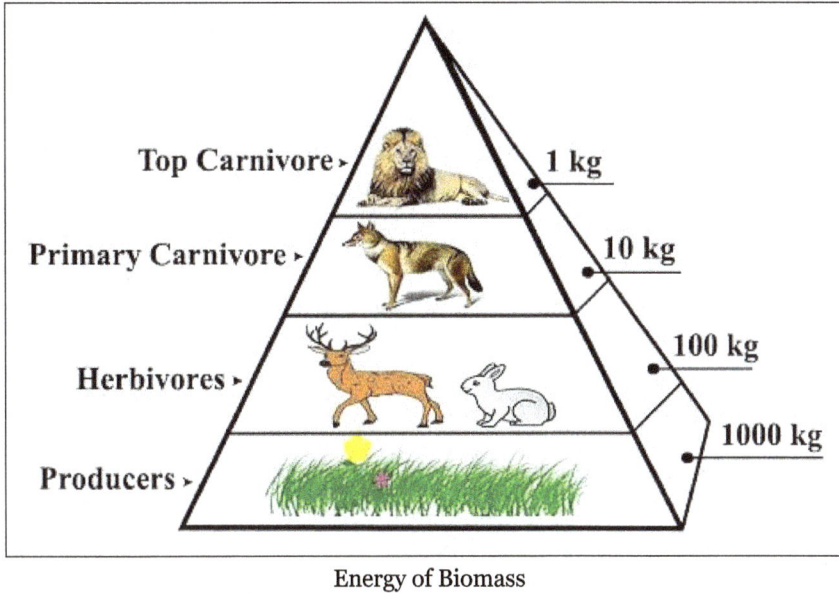

Energy of Biomass

The above figure shows that there is a decrease in the biomass from the lower tropic level to the higher tropic level. This is because the producers (trees) are maximum in the forest, which contribute a huge biomass. The next tropic levels are herbivores (insects, birds) and carnivores (snakes, foxes). The top of the tropic level contains few tertiary consumers (lions and tigers), the biomass of which is very low.

1.6   Forest Ecosystem

The entire assemblage of organisms together with their environmental substrate, interacting inside a defined boundary. Forests and woodlands occupy about 38% of the Earth's surface and they are more productive and have greater biodiversity than other types of terrestrial vegetation.

The Ecological Benefits

Production of oxygen:

During photosynthesis trees produce oxygen which is essential for life on earth.

Reducing global warming:

The main greenhouse gas carbon dioxide (CO_2) is absorbed by the trees (forests).

The trees absorb the main greenhouse gas carbon dioxide (CO_2), which is a raw material for photosynthesis. Thus the problem of global warming causes by greenhouse gas CO_2 is reduced.

Soil conservation:

Roots of trees bind the soil tightly and prevent soil erosion. They also act as wind breaks.

Regulation of hydro logical cycle:

Watersheds in forest act like giant sponges which absorb rainfall, slow down the runoff and slowly release the water for recharge of springs.

Pollution modulators:

Forests can absorb many toxic gases and noises and helps in preventing air and noise pollution.

Wildlife Habitat:

Forests are the homes of millions of wild animals and plants.

Characteristics of Forest Ecosystems

Forests are characterized by warm temperature and adequate rainfall, which make the generation of number of ponds, lakes etc.

- The forest maintains climate and rainfall.
- The forest support many wild animals and protect biodiversity.
- The soil is rich in organic matter and nutrients, which support the growth of trees.
- Since penetration of light is so poor, the conversion of organic matter into nutrients is very fast.

Types of Forest Ecosystem

Depending upon the climate conditions, forest can be classified into the following types:

- Tropical deciduous forests.
- Tropical rain forests.
- Temperate rain forests.
- Tropical scrub forests.
- Temperate deciduous forests.

Features of Different Types of Forests

Tropical Rain Forests:

They are found near the equator. They are characterized by high temperature. They have broad leaf trees like teak and sandal and the animals like lion, tiger and monkey.

Tropical Deciduous Forests:

They are found little away from the equator. They are characterized by a warm climate and rain is only during monsoon. They have different types of deciduous trees like maple, oak and animals like deer, fox, rabbit and rat.

Tropical Scrub Forests:

These are characterized by a dry, climate for longer time. They have small deciduous trees and shrubs and animals like deer, fox, etc.

Temperate Rain Forests:

They are found in temperate areas with adequate rainfall. They are characterized by coniferous trees like pines, red wood, firs, etc. and animals like squirrels, fox, cats, bear, etc.

Temperate Deciduous Forests:

They are found in areas with moderate temperatures. They have major trees including broad leaf deciduous trees like oak, hickory and animals like deer, fox, bear, etc.

Structure and Function of Forest Ecosystem

Abiotic Components

The abiotic components are inorganic and organic substances found in the soil and atmosphere. In addition to minerals, the occurrence of litter is characteristic features of majority of forests.

Examples: Climatic factors and minerals.

Biotic Components

Producers:

The plants absorb sunlight and produce food through photosynthesis.

Examples: Trees, Shrubs and ground vegetation.

Consumers:

i. Primary Consumers

They directly depend on the plants for their food.

Examples: Ants, flies, insects, mice, deer, squirrels.

ii. Secondary Consumers

They directly depends on the herbivores for their food.

Examples: Snakes, birds, fox.

iii. Tertiary Consumers

They depend on the primary carnivores for their food.

Examples: Animals like tiger, lion, etc.

Decomposers:

They decompose the dead plant and animal matter. Rate of decomposition in tropical and subtropical forests is more rapid than in the temperate forests.

Examples: Bacteria and fungi.

1.6.1 Grassland Ecosystem

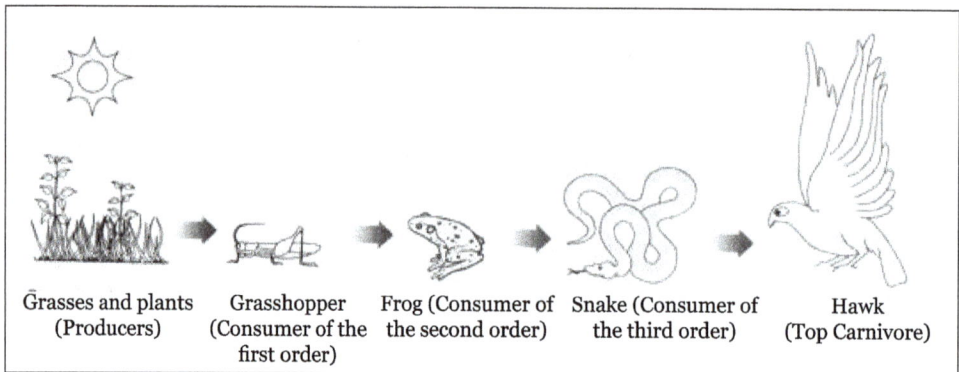

| Grasses and plants (Producers) | Grasshopper (Consumer of the first order) | Frog (Consumer of the second order) | Snake (Consumer of the third order) | Hawk (Top Carnivore) |

Food chain in grassland ecosystem.

Grasslands occupy about 19% of the earth's surface. The major grassland ecosystems of the world are the great plains of Canada and United States, Argentina to Brazil and Asia to Central Asia.

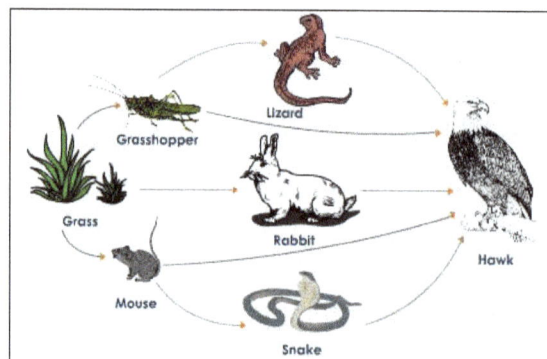

Food web in a grassland ecosystem.

The various components of a grassland ecosystem are as follows:

Abiotic substances:

These include the nutrients present in the soil and the aerial environment. The elements required by plants are hydrogen, oxygen, nitrogen, phosphorous and sulphur. These are supplied by the soil and air in the form of CO_2, water, nitrates, phosphates and sulphates. In addition to these some trace elements are also present in the soil.

Primary producers:

These are mainly grasses of the family, Graminae, large variety of herbs, some shrubs and scattered trees.

Consumers:

Herbivores such as grazing mammals, insects, some termites and millipedes are primary consumers.

The animals like fox, jackals, snakes, frogs, lizard, birds etc., are the carnivores feeding on herbivores. These are the secondary consumers of the grassland ecosystem. Hawks occupy the tertiary trophic level as these feed on the secondary consumers.

Decomposers:

These include bacteria, molds and fungi. These bring the minerals back to the soil to be available to the producers again.

Types of Grasslands

The tropical Savannah and Temperate Grassland are largely distinguished by differences in the temperature and the rainfall, both critical elements to grassland's formation. An area that receives very little rain becomes a desert; an area that receives significant amounts of rain often develops into forest. Grasslands hang somewhere in the balance.

Tropical Savannah's: It is found in Africa, Australia, South America and Indonesia, stay warm all year. They receive 50 to 130 cm during the rainy season (6 to 8 months) and endure drought for the remainder of the year. Plant and animal species vary greatly across Savannah, curbed by differences in climate, but much of the Savannah is characterized by thin soil where only grasses and flowering plants can grow. Like Canada's grasslands, this ecosystem supports an astonishing diversity of species, the African savanna, for example, is home to some of the world's most iconic mammals, including giraffes, zebras and lions.

Temperate Grasslands: It include Canadian grassland ecosystems, are also found around the globe. The plant and animal species in temperate grasslands are shaped by less rainfall (25 to 90 cm) and cycle through a greater range of seasonal temperatures. Many temperate grassland animals, which must adapt to dry, windy conditions, are recognizable to Canadians: grazing species like antelope and elk, burrowing animals like prairie dogs and badgers and predators like snakes and coyotes.

The dramatic contours of Canada's grasslands are the result of glacial movement and melting ice, which shaped this landscape over the last two hundred million years. Grasslands National Park, for example, boasts glacial melt water channels that feature plateaus, coulees, buttes that rise abruptly at horizon and layers of rock formation that hold fossilized secrets from 80 million years ago.

Characteristics of the Grassland

Precipitation

Grasslands make up 25 percent of the Earth's land surface and dominate in regions with limited rainfall, which prevents forest growth. This is the result of nearby mountain ranges that cause rain shadows over adjacent open range lands. Usually, grasslands have not only limited but also unpredictable rainfall and droughts are common. Where rainfall is even less, deserts will form. Savannas, on average, receive roughly 76 to 101 centimeters (30 to 40 inches) of rain per year, but steppes only average 25 to 51 cm (10 to 20 inches) per year. Prairies tend to be intermediate between savannas and steppes with 51 to 89 centimeters (20 to 35 inches) per year.

Temperature

It very much more in temperate grasslands than they do in savannas. Savannas are located in warm climates with average annual temperatures that only vary between 21 and 26 degrees Celsius (70 and 78 degrees Fahrenheit). They usually have only two seasons, a wet and a dry season. Temperate grasslands are characterized by hot summers where temperature can exceed 38 degrees Celsius (100 degrees Fahrenheit) and cold winters that can drop below negative 40 degrees Celsius (negative 40 degrees Fahrenheit).

Fire

Fires are important grassland characteristic. Regular fires promote the growth of native grasses but limit the growth of trees. Native grasses have deeper root systems that can survive fires, but invasive plants tend to have shallower roots and succumb to fires. Development has curtailed the number and extent of grassland fires and the lack of seasonal fires threaten the health of the world's grasslands. As of 2013, only 5% of the world's grasslands were protected and properly maintained and they remain the most endangered biome in the world.

Flora and Fauna

Savannas are home to some of the largest mammals on the planet like elephants, giraffes, rhinos, lions and zebras. Temperate grasslands are also home to large mammals, particularly bison and horses, medium-sized mammals like deer, antelope and coyotes, as well as small mammals such as mice and jack rabbits. The type of grasses that grow depend upon the amount of rainfall. Shorter steppe grasses often consist of buffalo grass and savanna grasses will contain taller grasses like bluestem and rye.

1.6.2 Desert Ecosystem

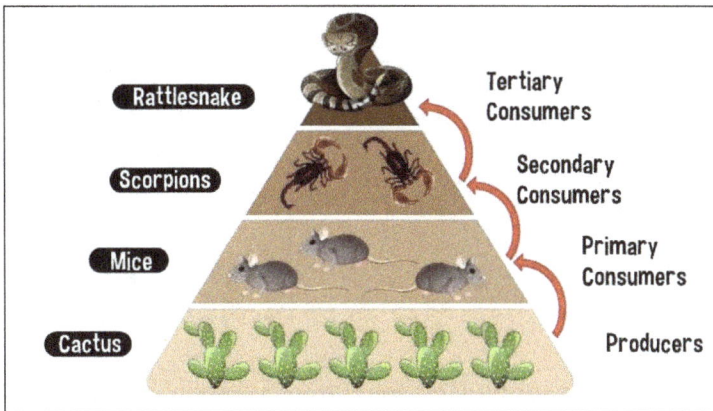

Desert Ecosystem.

One can find at least one desert on every continent except Europe and Antarctica. In order to be considered as a desert, it must receive less than 10 inches of water per year.

There are plenty of differences between the deserts of world. Some deserts are made of fine, red sand, others consist of sand mixed with pebbles and rocks. The desert sand started out as rock but after some years, weathering by wind and water has created dunes in the deserts. These sands are mostly minerals and sometimes oil can be found hidden deep within the rocks.

Types and Characteristic Features

One can find at least one desert on every continent except Europe and Antarctica. Each desert is different in some way, but they all have one thing in common. In order for an area of land to be considered a desert, it must receive less than 10 inches of water a year.

Clouds are scarce in these regions and we all know that without the clouds, there can't be rain, snow or any other precipitation. But clouds also serve another purpose, they block out some of Sun. The desert gets mighty hot during the day because the Sun beats down on the sand. At night, desert gets very cold, because there aren't clouds around to keep the heat from escaping to the atmosphere.

There are plenty of differences between the deserts of the world. Some deserts are made of very fine, red sand; others consist of sand mixed with pebbles and rocks. The desert sand started out as rock, but years of weathering by wind and water has created dunes in the deserts. These sands are mostly minerals and sometimes oil can be found hidden deep within the rocks.

Structure and Function

The various components of a desert ecosystem are:

Abiotic Component:

It includes the nutrients present in the soil and in aerial environment. The characteristic feature of the abiotic component is the lack of organic matter in the soil and scarcity of water.

Biotic Component:

The various biotic components representing three functional groups are:

i. Producer organisms

The producers are mainly shrubs or bushes, some grasses and a few trees. Surprisingly, there are many species of plants that can survive in the desert. Most of them are succulents, which mean they store water. Others have the seeds that lay dormant until a rain awakens them. Regardless, these plants find a way to get water and protect them from the heat.

The most famous desert plant is cactus. There are many species of cacti. The saguaro cactus is the tall, pole shaped cactus. The saguaro can grow to a height of 40 feet. It can hold several tons of water inside its soft tissue. Like all other cacti, saguaro has a thick, waxy layer that protects it from the Sun.

Other succulents include the desert rose and the living rock. This strange plant looks like a spiny rock. Its disguise protects it from predators. The welwitschia is a type of weird looking plant. It has two long leaves and a big root. This plant is actually a type of tree and it can live for thousands of years.

There are many other kinds of desert plants. Some of them have beautiful flowers; others have thorns and deadly poisons. Even in worst conditions, these plants continue to thrive.

ii. Consumers

These include animals such as the insects and reptiles. Besides them, some rodents, birds and some mammalian vertebrates are also found.

Desert Insects and Arachnids:

There are plenty of insects in the desert. One of the most common and destructive pests is the locust. Locust is a special type of grasshopper. They travel from place to place, eating all the vegetation they find. It can destroy many crops in a single day.

Not all desert insects are bad, though. The yucca moth is very important to the yucca plant, because it carries pollen from the flower to the stigma. The darkling beetle has a hard, white, wing case that reflects the Sun's energy. This allows the bug to look for food during the day.

There are also several species of ants in desert. The harvester ants gather seeds and store them for use during the dry season. And the honey pot ants have a very weird habit. Some members of the colony eat large amounts of sugar, so much that their abdomens get too large for them to move. The rest of the colony feeds off this sugar.

There are also arachnids in the desert. The spiders are the most notable arachnids, but scorpions also belong in this group. Some species of scorpions have poison in their sharp tails. They sting their predators and their prey with the piercing tip.

Desert Reptiles:

Reptiles are some of the most interesting creatures of the desert. Reptiles can withstand the extreme temperatures because they can control their body temperatures very easily. We can put most of the desert reptiles into one of two categories: snakes and lizards.

Many species of rattlesnakes can be found in the desert. Rattlesnakes have a noisy rattle they use to warn enemies to stay away. If the predator is not careful, the rattlesnake will strike, injecting venom with its sharp fangs. Other desert snakes include cobra, king snake and the hognose.

Lizards make up the second category of desert reptiles. They are probably the most bizarre looking animals in the desert. While some change colors and have sharp scales for defense, others change their appearance to look more threatening.

One such creature is the frilled lizard. When enemies are near, lizard opens its mouth, unveiling a wide frill. This makes the lizard look bigger and scarier. The shingle back has a tail with the same shape as its head. When a predator bites at the tail, the shingle back turns around and bites back. There are only two venomous lizards in the world and one of them is the gila monster. It has a very painful bite.

Desert Birds:

Like the other inhabitants of desert, birds come up with interesting ways to survive in the harsh climate. The sand grouse has special feathers that soak up water. It can then carry the water to its young trapped in the nest.

Other birds, like gila woodpecker, depend on the giant saguaro as its home. This woodpecker hollows out a hole in the cactus for a nest. The cool, damp inside is safe for the babies.

The roadrunner is probably the most well-known desert bird. The roadrunners are so named because they prefer to run rather than fly. Ostriches also prefer to use their feet. Even the young depend on walking to find food and water. The galah is one of the prettiest desert birds. It is one of the few species that return to the same nest year after year.

Galahs are interesting birds, in that the number of eggs they lay depends on climate. If the desert is in a drought, they don't lay any. However, during more tolerable years, the galah may lay as many as five eggs.

Desert Mammals:

There are several species of mammals in the desert. They range in size from a few inches to several feet in length. Like other desert wildlife, mammals have to find ways to stay cool and drink plenty of water. Many desert mammals are burrowers.

They dig holes in the ground and stay there during hot days. They return to the surface at night to feed. Hamsters, rats and their relatives are all burrowers. Not only do the burrows keep the animals cool, they are also a great place to store food.

Of course, not all animals have in holes in the ground. Kangaroo and spiny anteater both live in the Australian desert region. Spiny anteaters are unusual mammals because they lay eggs.

The desert is also full of wild horses, foxes and jackals, which are part of canine family. And we can't forget the cats. Lions are found all over the deserts of southern Africa. They get their water from blood of their prey.

Camels: The Cars of the Desert:

Camels could be included in the mammal section. Camels are the cars of the desert. Without them, people would have great difficulty crossing the hot terrain. There are two types of camels: Bactrian and dromedary. The main difference between the two is the number of humps. The Dromedaries have one hump and Bactrian have two. Both kinds are used by people, but only Bactrian's are found in the wild.

The camels are great for transportation because they use very little water. Camels can withstand very high temperatures without sweating. They also store fat in their humps for food. If a Bactrian camel travels a long distance without eating, its hump will actually get smaller.

iii. Decomposers

Due to poor vegetation the amount of dead organic matter is very less. As a result the decomposers are very few. The common decomposers are some bacteria and fungi, most of which are thermophile.

1.6.3 Aquatic Ecosystems

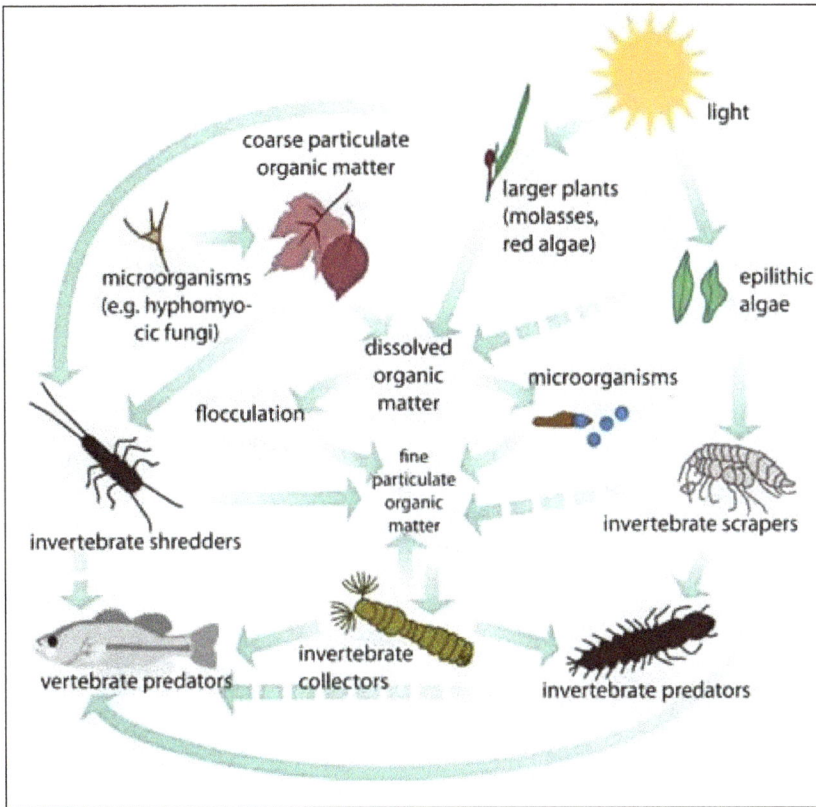

coarse particulate organic matter

larger plants (molasses, red algae)

light

microorganisms (e.g. hyphomyo- cic fungi)

epilithic algae

dissolved organic matter

flocculation

microorganisms

fine particulate organic matter

invertebrate shredders

invertebrate scrapers

vertebrate predators

invertebrate collectors

invertebrate predators

Aquatic Ecosystem.

The aquatic ecosystem deals with water bodies. The major types of organisms found in aquatic environments are determined by the water's salinity.

Types of Aquatic Life Zone

Aquatic life zones are divided into two types:

- Fresh water life zone.

Example: Ponds, streams, lakes, rivers.

- Salt water life zone.

Example: Oceans, estuarine.

Pond Ecosystem

A pond is a fresh water aquatic ecosystem where water is stagnant. It receives enough water during rainy season. It contains several types of algae, aquatic plants, insects, fishes and birds.

Characteristics

- Pond is temporary, only seasonal.

- It is stagnant fresh water body.

- Ponds get polluted easily due to limited amount of water.

Structure and Function

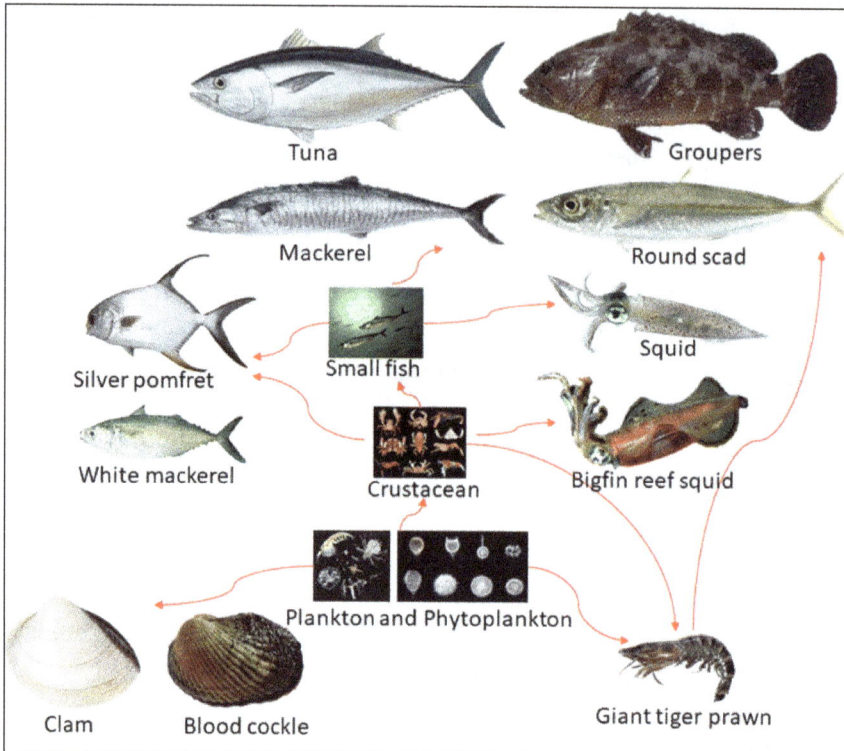

Marine ecosystem.

Abiotic components:

Ex: temperature, light, water and organic and inorganic compounds.

Biotic components:

i. Producers

They include green photosynthesis organisms. They are of two types:

a. Phytoplankton

There are microscopic aquatic plants, which freely float on the surface of water.

Example: Algae and small floating plants like volvox.

b. Microphytes

Example: Large floating plants and submerged plants.

ii. Consumers

a. Primary Consumers (Zooplanktons)

There are microscopic organisms which freely float on the surface of water.

Example: Protozoa, very small fish, flagellates and protozoans.

Zooplanktons are found with Phytoplankton. They feed on plants.

b. Secondary Consumers (Carnivores)

Example: Insects like water beetles and small fish.

They feed on zoo-plankton.

c. Tertiary Consumers

Example: Large fish like game fish.

They feed on smaller fish.

iii. Decomposers

Example: Fungi, bacteria and flagellates.

They decompose the dead plant and animal matter and their nutrients are released and secured by green plants.

Chapter 2

Natural Resources

2.1   Natural Resources and its Associated Problems

Continuous increase in the population caused an increasing demand for natural resources. Due to the urban expansion, electricity need and industrialization, man started utilizing natural resources at a much larger scale. Non-renewable resources are limited.

After some time, these resources may come to an end. It is a matter of much concern and ensures a balance between population growth and utilization of resources. This over utilization creates many problems. In some regions, there are problems of water logging due to over irrigation.

In some areas, there is no sufficient water for industry and agriculture. Thus, there is need for conservation of natural resources.

There are many problems associated with natural resources:

Forest resources and associated problems:

- Timber extraction.

- Deforestation.

- Dams and their effects on forests and tribal people.

- Use and over exploitation.

- Mining and its effects on forest.

Water resources and associated problems:

- Floods, droughts, etc.

- Use and overutilization of water.

- Dams and problems.

- Conflicts over water.

Mineral resource and associated problems:

- Environmental effects of extracting and using minerals.

- Use and exploitation.

Food resources and associated problems:

- Water logging and salinity.

- Changes caused by agriculture and over grazing.

- Effects of modern agriculture.

- World food problems.

- Fertilizer pesticide problems.

Energy resources and associated problems:

- Growing energy needs.

Land resources and associated problems:

- Land degradation.

- Soil erosion and desertification.

- Man-induced landslides.

2.1.1 Forest Resources: Use and Over-exploitation

Forest Resources

Forest, an important resource to mankind, supports in a direct or indirect way the development of human economy and society with its thousands of material or non-material products. The forest is also the main object for the continental ecological system.

The successful management of forest inheritance plays a vital role in managing the global environment, conserving the other related resources and guaranteeing the sustainable development of economy and society.

Over-exploitation

Due to overcrowded population materials supplied by the forest like Medicines, Shelter and Fuel are not sufficient to meet the people's demand. Thus the exploitation of forest materials is going on increasing day by day.

With growing civilization, the demand for raw materials like timber, pulp, minerals, fuel wood, etc., increases resulting in large scale logging, mining, road building and deforestations.

Reason for Over Exploitation in India

It has been estimated that in India the minimum density of forests required to maintain good ecological balance is about 33, all area. But, at present it is only about 22%. So over exploitation forest materials occur.

Causes in Over Exploitation

Over exploitation of forest wealth in the developing countries occurs in the following ways:

- Increasing agricultural production.
- Increasing industrial activities.
- Increasing in demand of wood resources.

2.1.2 Deforestation, Timber Extraction and Mining

Deforestation is the process of elimination or removal of forest resources due to many natural or man-made activities. In general, deforestation means destruction of forests.

Causes of Deforestation

Developmental Projects

Developmental projects cause deforestation in two ways:

- Through submergence of forest area under water.
- Destruction of forest area.

Example: Hydroelectric projects, big dams, road constructions, etc. Hence, there is a need to discourage the undertaking of any development works in the forest area.

Mining Operations: Mining have a serious impact on forest areas. Mining operation reduces the forest area.

Example: Mica, Coal, Limestone, etc.

Raw Materials for Industries: Wood is the important raw material for so many purposes.

Example: For making boxes, ply woods, etc.

Fuel Requirements: In India, both tribal and rural population is dependent on the forest for meeting their daily needs of fuel wood, which leads to the pressures on forest, ultimately to deforestation.

Shifting Cultivation: The replacement of natural forest ecosystem for mono specific tree plantation can lead to disappearance of number of plant and animal species.

Forest Fires: Forest fire is one of the major causes for deforestation. Due to human interruption and rise in ambient temperature.

Effects of Deforestation on the Investment

Since many people are dependent on the forest resources, deforestation will have the following- social, ecological and economic effects.

Soil Erosion:

Deforestation also causes soil erosion, drought, landslides, and floods. Natural vegetation acts as a natural barrier to reduce the wind velocity, which in turn reduces soil erosion.

Loss of Biodiversity:

Most of the species are very sensitive to any disturbance and changes. When the plants no longer exist, animals depend on them for food and habitat become extinct.

Loss of Food Grains:

As a result of soil erosion, the countries loose the food grains.

Timber Extraction and Mining

Timber extraction, dams and mining are invariable parts of the needs of a developing country. If timber is overharvested the ecological functions of forest are lost. Unfortunately forests are located in areas where there are rich mineral resources.

Forests also cover the steep embankments of river valleys, which are ideally suited to develop hydel and irrigation projects. Thus a constant conflict of interests between the conservation interests of the environmental scientists and the Mining and Irrigation Departments is present. What needs to be understood is that long term ecological gains cannot be sacrificed for short-term economic gains that unfortunately lead to deforestation. These forests where development projects are planned and can displace thousands of tribal people who lose their homes when these plans are executed.

Timber Extraction

There has been unlimited exploitation of timber for commercial use. Due to increased industrial demand, timber extraction has significant effect on forest and tribal people.

Logging

- Poor logging results in degraded forest and may leads to soil erosion especially on slopes.

- New logging roads permit shifting cultivators and fuel wood gatherers to gain access to the logging area.

- Loss of long term forest productivity.

- Species of plants and animals may be eliminated.

- Exploitation of tribal people by contractor.

Mining

Mining from shallow deposits is done by surface mining while that from deep deposits is done by sub-surface mining. It leads to degradation of lands and loss of top soil. It is estimated that about eighty thousands hectare land is under stress of mining activities in India.

Mining leads to drying up perennial sources of water sources like streams and spring in mountainous area. Mining and other associated activities remove vegetation along with underlying soil mantle, which results in destruction of topography and landscape in the area. Large scale deforestation has been reported in Mussorie and Dehradun valley due to indiscriminating mining.

The forested area has declined at an average rate of about 33% and the increase in non-forest area due to mining activities has resulted in relatively unstable zones leading to landslides.

Indiscriminate mining in forests of Goa since 1961 has destroyed more than 50000 ha of forest land. The coal mining in Jharia, Raniganj and Singrauli areas has caused extensive deforestation in Jharkhand.

Mining of magnetite and soapstone have destroyed 14ha of forest in hilly slopes of Khirakot, Kosi valley and Almora.

Mining of the radioactive minerals in Kerala, Tamilnadu and Karnataka are posing similar threats of deforestation.

The rich forests of Western Ghats are also facing the same threat due to mining projects for excavation of copper, chromites, bauxite and magnetite.

2.1.3 Dams and other Effects on Forest and Tribal People

Pandit Jawaharlal Nehru referred dam and valley projects as "Temples of modern India". These big dams and rivers valley projects have multipurpose uses. However, these dams are also responsible for the destruction of forests. They are responsible for loss of flora and fauna, degradation of catchment areas, disturbance in forest ecosystems and increase of water borne diseases, rehabilitation and resettlement of tribal peoples.

India has more than 1550 large dams, the maximum being in the state of Maharashtra, followed by Gujarat and Madhya Pradesh.

The highest one is Tehri dam, on river Bhagirathi in Uttaranchal and the largest in terms of capacity is Bhakra dam on river Satluj in Himachal Pradesh. Big dams have been in sharp focus of various environmental groups all over world, which is mainly because of several ecological problems including deforestation and socioeconomic problems related to tribal or native people associated with them.

The Silent valley hydroelectric project was one of the first such projects situated in the tropical rain forest area of Western Ghats which attracted much concern of the people.

The crusade against ecological damage and deforestation caused due to Tehri dam was led by Shri. Sunder Lal Bahaguna, the leader of Chipko Movement.

The cause of Sardar Sarovar Dam related issues have been taken up by the environmental activitist Medha Patkar, joined by Arundhati Ray and Baba Amte. For building big dams, large scale devastation of forests takes place which breaks the natural ecological balance of the region.

Droughts, floods and landslides become more prevalent in such areas. Forests are the repositories of invaluable gifts of nature in the form of biodiversity and by destroying them; we are going to lose these species even before knowing them.

These species could be having marvelous economic or medicinal value and deforestation results in loss of this storehouse of species which have evolved over millions of years in a single stroke.

Effects of Dams

Dams are massive artificial structures built across the river to create a reservoir in order to store water for many beneficial purposes. However, these dams are also responsible for the destruction of forest and displacement of local people.

The Indian Scenario

India has more than 1600 large dams. Dam is the highest built across the river in the State.

State	No. of Dams
Maharashtra	More than 500 dams
Gujarat	More than 250 dams
Pradesh	More than 130 dams

Effects of Dam on Forest:

- Thousands of hectares of forest have been cleared for executing river valley projects.

- In addition to the dam construction, the forest is also cleared for residential accommodation, office buildings, etc.

- The big river valley projects also cause water logging which leads to salinity and in turn reduces the fertility of the land.

- Hydroelectric projects also have led to widespread loss of forest in recent years.

- Hydroelectric projects provide opportunities for the spread of water borne diseases.

- Construction of dams under these projects leads to killing of wild animals and destroying aquatic life.

Examples:

- Narmada Sugar Project: It has to submerged 3.5 each hectares of forest comprising of teak and bamboo trees.

- Dam: It has submerged 1000 hectares of forest affecting about 430 species of plants.

Effects of dam on tribal people:

- The greatest social cost of big dam is the widespread displacement of tribal people such that biodiversity cannot be tolerated.

- The displacement and cultural change affects the tribal people both mentally and physically. They do not accommodate modern food habits and life styles.

- Tribal people and their culture cannot be questioned and destroyed.

- Generally, the body conditions of the tribal people will not suit with the new areas and hence they will be affected by many diseases.

Conflicts of Water Resources

Conflicts through use:

- Unequal distribution of water led to interstate or international disputes.

- International conflicts: India and Pakistan fight to get water from the Indus.

Construction of Dams/Power stations:

For hydroelectric power generation, dams built across the rivers, initiates conflict between the states.

- Conflict through pollution.

- Rivers and Lakes are used for electricity, shipping and for industrial purpose.

- Disposal of waste water and industrial waste decrease the quality of water and causes pollution.

2.2 Use and Over Utilization of Surface and Ground Water

Water Resources

The water cycle, through evaporation and precipitation, maintains hydrological systems which form rivers and lakes and support in a variety of aquatic ecosystems. The wetlands are intermediate forms between the terrestrial and aquatic ecosystems and contain species of plants and animals that are highly moisture dependent. The aquatic ecosystems are used by several people for their daily needs such as drinking water, cooking, washing, watering animals and irrigating fields.

The world depends on a limited quantity of fresh water. Water covers 70% of the surface of earth but only 3% of this is fresh water. Of this, 2% is in polar ice caps and the remaining 1% is usable water in rivers, lakes and subsoil aquifers. Only a fraction of this can be actually used.

At a global level, 70% of water is used for agriculture about 25% for industry and only 5% for domestic use. Although this may varies in different countries, industrialized countries use a greater percentage for industry.

India uses 90% for agriculture, 7% for industry and 3% for domestic use. One of the greatest challenges facing the world in this century is the need to rethink the overall management of water resources.

Main sources of water are:

- Surface water sources: Lakes impounding reservoirs, streams, seas, irrigation canals.

- Ground water sources: Springs, wells, infiltration wells.

Merits of ground water sources:

- The water quality is good and better than surface source.

- Being underground, the ground water supply has less chance of being contaminated by atmospheric pollution.

- The land above ground water source can be used for other purposes and has less environmental impacts.

- Prevention of water through evaporation is ensured and thus loss of water is reduced.

- Ground water supply is available and can even be maintained in deserted areas.

Demerits of ground water source:

- Modeling, analysis and calculation of ground water is less reliable and based on the past experience, thus posing high risk of uncertainty.

- Water obtained from ground water source is always pressure less. A pump is required to take the water out and is then again pumped for daily use.

- The transport/transmission of ground water is a problem and an expensive work. Water has to be surfaced or underground conduits are required.

- Boring and excavation for finding and using ground water is expensive work.

Impounding Reservoirs

It is a basin constructed in the valley of a stream or river for the purpose of holding stream flow so that the stored water may be used when supply is insufficient.

They have the following two functions:

- To impound water for beneficial use.

- To retard flood.

Two functions may be combined to some extent by careful operations.

An impounding reservoir presents a water surface for evaporation. This loss must be considered. Possibility of large seepage loss must also be considered. If it is economically impossible to prevent them, the project may have to be abandoned or move it to a more favorable site. There will be some loss by seepage through and under the dam itself.

Well Hydraulics

Fresh water is a natural and renewable resource. Rate at which it is renewed is becoming surpassed by the rate that is being consumed in the Tampa Bay area. Job of finding new sources of fresh water and managing the existing sources have fall under authority of the Southwest Florida Water Management District.

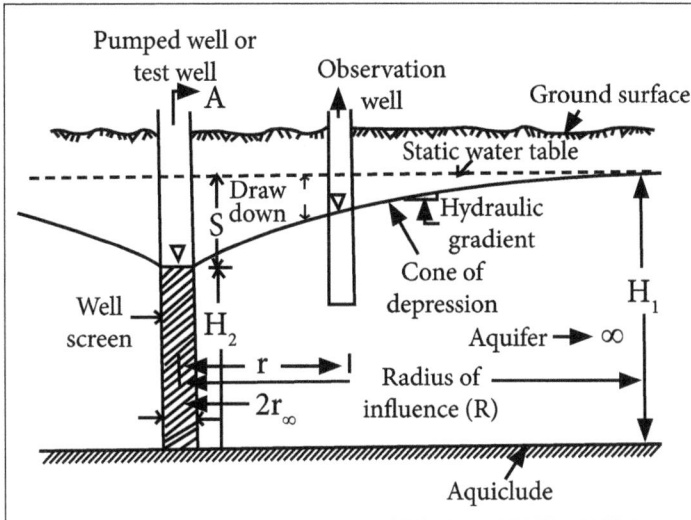

Well Hydraulics.

Agriculture needs of water for the Crops:

As our country is an agricultural based country, the crops are to be developed for the production of different types of foods grains. The requirement of water varies from crop to crop. Different research stations are busy in identifying the water needs of all the crops. The most of these crops are shallow rooted, thus water being extracted from to top layers of the soil. The soil moisture available in the top layers is essential for such crops. Hence, it is advocated to keep the topsoil always moist so that crops do not wilt under no water of condition.

The many scientific methods of ploughing are developed to maintain the moisture in the topsoil for longer periods.

The requirement of maintaining the water storage for breeding the fish and prawn culture is attaining importance whenever we go for storage project. If since fish cannot travel against flow when the velocity is high, special structures are provided adjacent to the weirs to maintain low velocities. Such structures are known as fish ladders.

The different purposes of the projects have their conflicting effect on the environment. However, depending on the priority of water needs of the area to be served, planning has to proceed in a way such that the impact on environment is kept to be minimum while serving the urgent and primary needs of the people of that area.

Groundwater and Aquifers:

The water that is available in the deeper layers of the earth is known as Groundwater. This water has been trapped inside earth's crust for several centuries. The water that is lying under the ground has the capacity to move in general direction of its slope with a very small velocity.

The bodies that contain such water are known as Aquifers. If water from these aquifers can be drawn by digging wells and pumping water from these wells. Some of these wells can supply very large quantities of water. In some cases, the drinking water needs of cities are completely met by groundwater bodies. In the agriculture sector also, groundwater is supporting the growth of crops enormously in certain agricultural dominated areas.

Groundwater and Wells:

The wells are used to bring groundwater to the land surface by means of pumps. If wells can be deep wells and shallow wells depending upon the depth at which ground water is available. If sometimes open dug wells are used where the water table is high.

In the case of deeper and hard rock aquifers, Tube wells are constructed. In such cases, deep well pumps such as turbine pumps or jet pumps are used to lift water to the surface of the ground. These tube wells in many cases yield significantly high discharges.

In the way, groundwater is the water that is available at the place of its use. Thus this water can be lifted from the wells in the agricultural fields and supplied to the crops raised in that field. Hence, there is no need to send to water long distances. This incidentally reduces the water losses considerably. The use of ground water for nearby areas also does not pose the problems of environmental degradation.

2.2.1 Floods

Floods have been a serious environmental hazard for a century. However, the havoc raised by rivers overflowing their banks has become progressively more damaging, as people have deforested catchments and intensified use of river flood plains that once acted as safety valves. Wetlands in the flood plains are nature's flood control systems into which the overfilled rivers could spill and act as a temporary sponge holding water and preventing fast flowing water from damaging the surrounding land. Deforestation in the Himalayas causes floods that year after year kill people, damage crops and destroy homes in the Ganges and its tributaries and the Brahmaputra.

Rivers can change their course during floods and tons of valuable soil is lost to the sea. As the forests are degraded, rainwater not a longer percolates that slowly into the subsoil but, runs off down in the mountainside bearing large amounts of topsoil. This blocks the rivers temporarily but gives way as the pressure mounts allowing enormous quantities of water to wash down suddenly into the plains below. There, rivers swell, burst their banks and flood waters spread to engulf people's farms and homes.

Cause and Effect of Floods

Whenever the magnitude of water flow exceeds the carrying capacity of channel within the banks, the excess of water overflows on the surroundings causes floods.

Causes of Floods:

- Heavy rain, rainfall during cyclone causes floods.

- Sudden snow melt also raises the quantity of water in streams and causes flood.

- Sudden and excess release of compounded water behind dams.

- Human activities like construction of roads, building and parking space that covers the earth's surface prevents infiltration.

- Clearing of forest for agriculture has also increases severity of floods.

Effects of Floods:

- Flood cause heavy suffering to people living in low lying areas because the houses and properties are washed away.

- Flood damage standing crops and livestock.

The unvarying and variable components of unvarying and variable characteristics:

	Stable Unvarying	Variable
Network Characteristics	Patters	Surface storage, under-drainage, channel length, contributing or source area etc.
Basin	Slope, attitude, shape, etc., of the basin	Arising from the interactions between climate, geology, soil type, vegetation cover, etc., which are manifested through the storage capacity of soil and bedrock, extent of infiltration and transmissibility of soil and bedrock that are affected by anthropogenic activities.
Channel Characteristics	Slope, flood control and river regulation works	Roughness, load, shape, storage.

Flood Management:

- Encroachment of flood ways should be banned.

- Building walls prevent spilling out. The flood water over flood plains.

- Diverting excess water through channels or canals to areas like lake, river, etc. where water is not sufficient.

- Build check dam on small streams, move building off the flood plains.

- Restore wetlands, replace ground course.

- River networking in the country also reduce flood.

- Flood forecast and flood warning are also given by the central water commission.

- Reduction of runoff by increasing infiltration through appropriate afforestation.

India Scenario-Floods

Next to Bangladesh, India is the most flood affected country in the world. Nearly 40 million hectares are affected by annual floods in India. 20% of which is present only in UR Next to UP, Orissa, Andhra Pradesh, Bihar.

2.2.2 Drought

The drought is due to lack or insufficiency of rain for an extended period that causes considerable hydrologic imbalances and consequently water shortages, stream flow reductions and depletion of groundwater levels and soil moisture. Drought is the most serious physical hazard to agriculture in nearly every part of the world.

If drought not only leads to serious economic consequences but also leaves behind untold human misery. Among all the natural disasters, drought affects largest number of people in the world. Shortage of water for even the basic needs is the main problem in the drought areas.

Even the shallow rooted crops do not grow in such areas. Getting sufficient drinking water is the another problem needing immediate attention in these areas. Some of the measures like infiltration wells, underground dams, small watersheds, are being taken up to alleviate the sufferings of the people residing in the drought prone areas.

Certain advance techniques such as Cloud Seeding and Artificial Rains are also being tried with varying successes. However, these methods are quite expensive and unpredictable in their success. Scant rains for extensive periods also lead to ecological changes. Ultimately, Government has found reasonable remedies in the form of development of small watersheds in such areas.

In case of most arid regions of the world, the rains are unpredictable. This leads to periods when there is a serious scarcity of water to drink, it use in farms or provide for an urban and industrial use. Drought prone areas are faced with irregular periods of famine. Agriculturists have no income in these bad years and as they have no steady income, they have a constant fear of droughts. India has 'Drought Prone Areas Development Programs' that are used in such areas to buffer the effects of droughts. Under these schemes, people are given wages in bad years to build the roads, minor irrigation works and plantation programs.

It is an unpredictable climatic condition and occurs due to the failure of one or more monsoons. It varies in frequency in different parts of our country. While it is not feasible to prevent the failure of the monsoon, good environmental management can reduce the ill effects. The scarcity of water during the drought affects homes, agriculture and industry. It also leads to the food shortages and malnutrition that especially affects children. Several measures can be taken to minimize the serious impacts of a drought.

However this should be done as a preventive measure so that if the monsoons fail, it's an impact on the life of the local people is minimized. In years when the monsoon is adequate, we use up the good supply of water without trying to conserve it and use the water judiciously. Thus during a year when the rains are poor, there is no water even for drinking in the drought area. One of the factors that worsen the effect of drought is deforestation.

Once the hill slopes are denuded of forest cover rainwater rushes down the rivers and is lost. Forest cover permits the water to be held in the area allowing it to seep into the ground. This charges the underground stores of water in natural aquifers. This can be used in drought years if the stores have been filled during a good monsoon. If water from an underground stores is overused the water table drops and vegetation suffers. This soil and water management and afforestation are long term measures which reduces the impact of drought.

2.2.3 Conflicts Over Water

Water conflict is a term describing a conflict between countries, states or groups over an access to water resources. The United Nations recognizes that water disputes result from opposing interests of water users, public or private.

A wide range of water conflicts appear throughout the history, though traditional wars waged over water alone. Instead, the water has historically been a source of tension and a factor in conflicts that start for other reasons. Although, water conflicts arise for several reasons, including territorial disputes, a fight for resources and strategic advantage. A comprehensive online database of water related conflicts the Water Conflict Chronology has been developed by Pacific Institute. This database lists violence over water which is going back nearly 5,000 years.

These conflicts occur over both the freshwater and saltwater and both between and within nations. However, conflicts occur mostly over freshwater, because the freshwater resources are necessary, yet limited, they are the center of water disputes arising out of need for potable water and irrigation. As freshwater is a vital unevenly distributed natural resource, its availability often impacts the living and economic conditions of a country or region.

The lack of cost-effective water supply options in areas like Middle East, among other elements of water crises can put severe pressures on all water users, whether the corporate, government or individual leading to tension and possible aggression. Recent humanitarian catastrophes, such as the Rwandan Genocide or the war in Sudanese Darfur, have been linked back to water conflicts.

A recent report "Water Cooperation for a Secure World" published by Strategic Foresight Group concludes that active water cooperation between countries reduces the risk of war. This conclusion is obtained after examining the trans-boundary water relations in over 200 shared river basins in 148 countries; a growing number of water conflicts are sub-national.

Sustainable Water Management

'Save water' campaigns are essential to make people everywhere aware of dangers of water scarcity. A number of measures are to be taken for the better management of the world's water resources. These include measures such as:

- Building several small reservoirs instead of few mega projects.

- Soil management, micro catchment development and afforestation permits recharging of underground aquifers thus reducing the need for large dams.

- Preventing loss in Municipal pipes.

- Treating and recycling municipal waste water for an agricultural use.

- Preventing leakages from dams and canals.

- Develop small catchment dams and protect wetlands.

- Effective rain water harvesting in urban environments.

- Pricing water at its real value makes people use it more responsibly and efficiently and reduces water wasting.

- Water conservation measures in agriculture namely using drip irrigation.

- In the deforested areas where land has been degraded, the soil management by bunding along the hill slopes and making 'nala' plugs, can help to retain moisture and make it possible to revegetate degraded areas. Managing a river system is best done by leaving its course as undisturbed as possible. The dams and canals lead to major floods in the monsoon and the drainage of wetlands seriously affects areas that get flooded when there is high rainfall.

2.2.4 Dams: Benefits and Problems

The dams are the major structures in any reservoir from the point of view of structural importance, design details and cost. The dams are of different types depending on different criteria. Depending on the material used for construction, dams can be: earthen dams, rock fill dams, masonry dams, concrete dams, steel dams and timber dams.

If based on the design, the dams can be: gravity dams, arch dams, buttress dams and multiple arch dams. Similarly, based on the purpose, the dams are known as overflow dams and non-overflow dams.

The masonry and concrete dams are more or less leak proof and hence seepage is not possible. Although in the case of earthen and rock fill dams, seepage of water is expected, in good quantities and therefore, possibilities of water logging on downstream side will be the adverse effect on the environment.

The water resources projects are constructed to many purposes depending on the needs of people of the area to be served. Whenever the projects are developed to supply water for various purposes, the projects are termed as multipurpose projects.

The different purposes can be: irrigation and agriculture, hydropower generation, flood control, navigation, drinking water supply, water for Industries, recreation and amusement parks and afforestation. All of the above purposes, irrigation and agriculture occupy higher priority, as the production of necessary food grains for one billion population of country is the primary concern to us.

To develop industries and other power needs, the next priority is the Hydropower development. The emphasis is increasing on the hydropower, as the natural resources for other forms of energy such as thermal are becoming scarce. Due to the rapid development of urban areas, scarcity of drinking water has surfaced in most of the cities. Hence, the present emphasis is on bringing water to the cities from storage reservoirs.

The slowly, water needs for drinking purposes is occupying the priority when compared to the other needs. In the flood plains, the problem of inundation of adjacent habituated areas is a priority, as during every flood the losses in respect of human and cattle life, crops, property and fertile soils, are bringing misery to people.

With every increasing movement of men and material, the transportation by navigation is also recognized as viable mode. In the recent times, water has been used for recreational purposes also. In some countries, water sports are gaining popularity. In order to develop greenery in many dry areas, Government is encouraging people to go in for tree plantations on a large scale resulting in afforestation.

Today there are more than 45,000 large dams around the world, which provide an important role in communities and economies that harness these water resources for their economic development. Current estimates suggest some 30-40% of irrigated land worldwide relies on dams. Hydropower, another contender for the use of stored water, currently supplies 19% of the world's total electric power supply and is used in over 150 countries. The world's two most populous countries China and India have built around 57% of the world's large dams.

Dams Problems

- Serious impacts on riverine ecosystems.

- Fragmentation and physical transformation of rivers.

- Dislodging animal populations, damaging their habitat and cutting off their migration routes.

- Water logging and salinization of surrounding lands.

- Fishing and travel by boat disrupted.

- Social consequences of large dams it's due to displacement of people.

- The emission of greenhouse gases from reservoirs due to rotting vegetation and carbon inflows from the catchment is a recently identified impact.

Large dams have had serious impacts on the livelihoods, lives, cultures and spiritual existence of indigenous and tribal peoples. They have suffered disproportionately from the negative impacts of the dams and often been excluded from sharing the benefits. In India, of 16 to 18 million people displaced by dams, 40 to 50% were tribal people, who account for only 8% of our nation's one billion people. Conflicts over dams have heightened in the last two decades because of their social and environmental impacts and failure to achieve the targets for sticking to their costs as well as achieving promised benefits. Recent examples show how failure to provide a transparent process that includes effective participation of the local people has prevented affected people from playing an active role in debating the pros and cons of the project and its alternatives. The loss of traditional, local controls over equitable distribution remains a major source of conflict.

2.3   Mineral Resources: Use and Exploitation

Minerals are naturally occurring substances having definite chemical composition and physical properties.

Formation of Mineral Deposits

Concentration of the mineral at a particular spot, which can be extracted profitably and gives rise to a mineral deposit. The formation of these deposits is a very slow biological process; it even takes millions of years to develop as a mineral deposit.

Various Biological Processes:

- Formation of mineral deposits is due to the biological decomposition of dead animals and organic matters.

- Mineral deposits also formed due to evaporation of sea water.

- Mineral deposits are formed due to oxidation reduction reaction inside the earth.

Classification of Mineral Resources

U.S. Geological Survey divides nonrenewable mineral resources into 3 categories:

- Identified resources.
- Undiscovered resources.
- Reserves.

Identified Resources:

The location, existence, quantity and quality of these mineral resources are known by the direct geological evidence and measurements.

Undiscovered Resources:

These mineral resources are assumed to be existing on the basis of geological knowledge and they but their locations, quantity and quality are unknown.

Reserves:

These mineral resources are identified resources, from which a usable minerals can be extracted profitably.

Uses of Minerals:

- Development of industrial plants and machinery.
- Construction, housing, settlements.
- Generation of energy.
- Communication purposes.
- Medicinal purposes, particularly in Ayurveda system.

Classification of Minerals

Minerals are classified into two ways based on this composition and usage:

Based on Composition:

Based on composition, minerals can be classified into two types:-

i. Metallic Minerals

Metallic minerals are the one from which various kinds of metals can be extracted.

Example: Iron, copper, zinc, etc.

ii. Non-Metallic Minerals

Nonmetallic minerals are the one from which various nonmetallic compound can be extracted.

Example: Quartz, folds par, calcite.

Based on Usage:

Based on usage, minerals are classified into two types:-

i. Critical Minerals

These are essential for the economic power of a country. Example: Iron, Al, Cu and Au.

ii. Strategic Minerals

These are required for the defense of a country.

Example: Manganese, cobalt.

Management of Mineral Resources

*   The efficient used and protection of mineral resource.
*   Modernization of the mining industries.
*   Search for new deposit.
*   Reuse and Recycling of the metals.
*   Environmental impacts can be minimized by adopting the ecofriendly mining technology.

Use and Exploitation

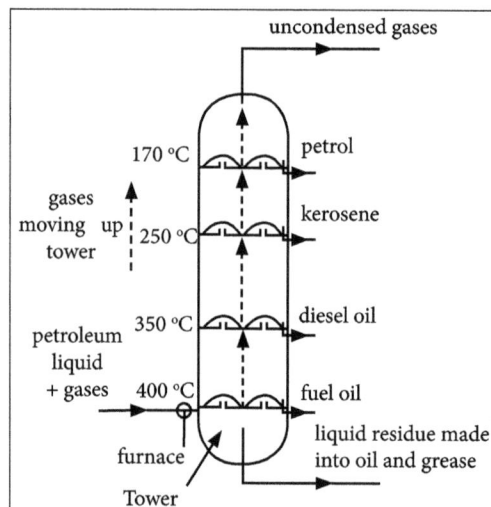

Fractional Distillation Tower.

The use of minerals varies greatly between countries. The greatest use of minerals occurs in developed countries. Like other natural resources, mineral deposits are unevenly distributed around on the earth. Some countries are rich in mineral deposits and other countries have no deposits. The use of the mineral depends on its properties. For example aluminum is light but strong and durable so it is used for aircraft, shipping and car industries.

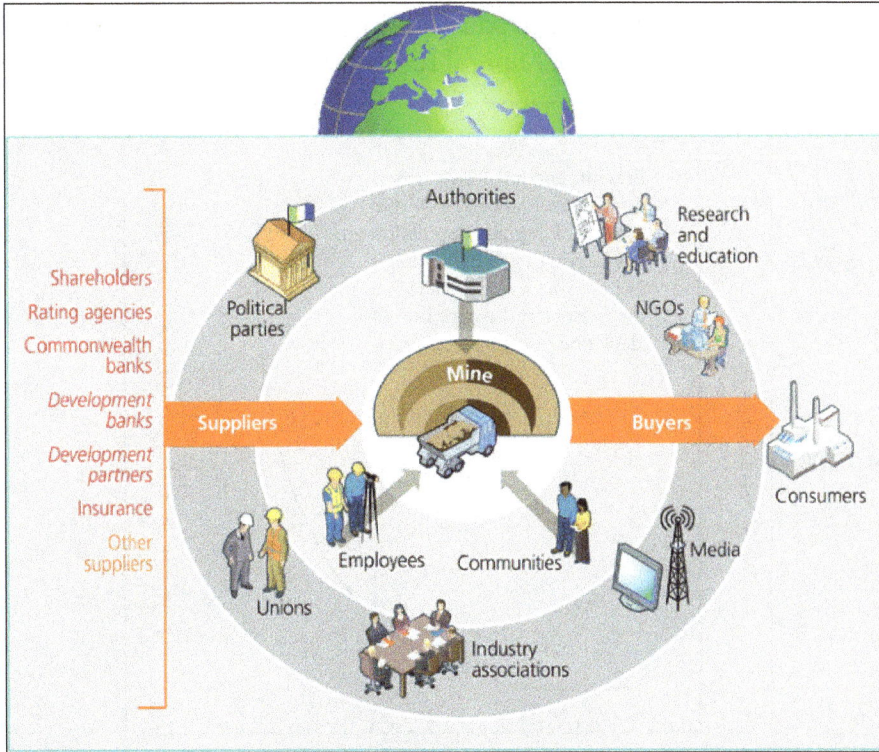

Uses of mineral resources.

Recovery of mineral resources has been with us for a long time. Early Paleolithic man found flint for arrowheads and clay for pottery before developing codes for warfare. And this was done without geologists for exploration, mining engineers for recovery or chemists for extraction techniques. Tin and copper mines were necessary for a Bronze Age, gold, silver and gemstones adorned the wealthy of early civilizations and iron mining introduced a new age of man.

Human wealth basically comes from the agriculture, manufacturing and mineral resources. Our complex modern society is built around the exploitation and use of mineral resources. Since the future of humanity depends on mineral resources, we must understand that these resources have limits; our known supply of minerals will be used up early in the third millennium of our calendar.

Additionally, modern agriculture and the ability to feed an overpopulated world are dependent on mineral resources to construct the machines that till the soil, enrich it with mineral fertilizers and to transport the products.

We are now reaching limits of reserves for many minerals. Human population growth and increased modern industry are depleting our available resources at increasing rates. The pressure of human growth upon the planet's resources is a very real problem.

The consumption of natural resources proceeded at a phenomenal rate during the past hundred years and population and production increases cannot continue without increasing pollution and depletion of mineral resources.

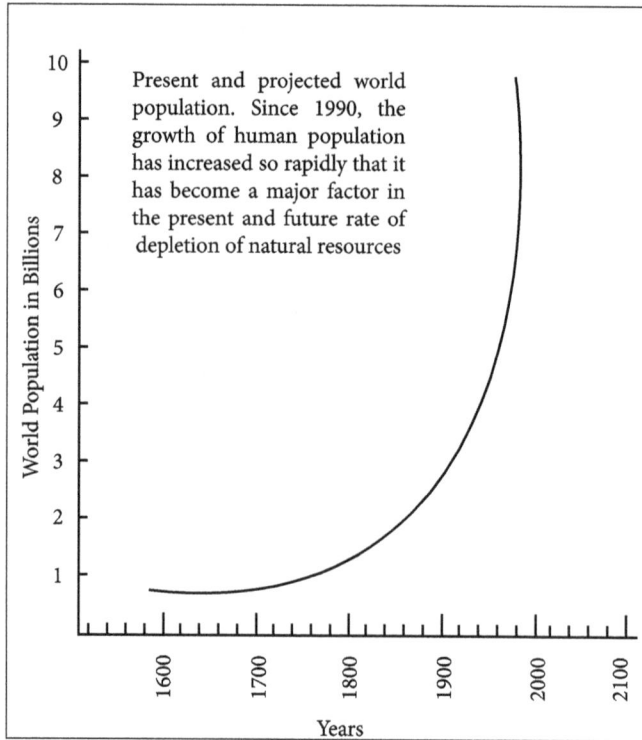

Present and projected world population. Since 1990, the growth of human population has increased so rapidly that it has become a major factor in the present and future rate of depletion of natural resources

Increase in World population.

The geometric rise of population as shown in Figure has been joined by a period of rapid industrialization, which has placed incredible pressure on natural resources. Limits of growth in the world are imposed not as much by pollution as by the depletion of natural resources.

As the industrialized nations of the world continue the rapid depletion of energy and mineral resources and resource-rich less-developed nations become increasingly aware of the value of their raw materials, resource driven conflicts will increase.

In Figure below, we see that by about the middle of next century the critical factors come together to impose a drastic population reduction by catastrophe. We can avert this only if we embark on a planet-wide program of transition to a new physical, economic and social world that recognizes limits of growth of both population and resource use.

In a world that has finite mineral resources, exponential growth and expanding consumption is impossible. Fundamental adjustments must be made to the present growth culture to a steady-state system.

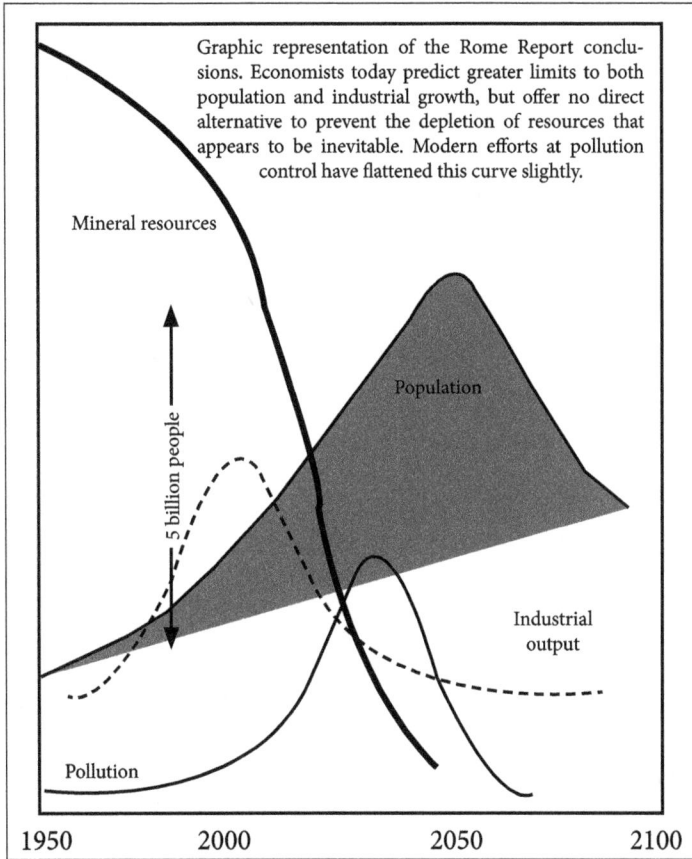

Graphic representation of the Rome Report conclusions. Economists today predict greater limits to both population and industrial growth, but offer no direct alternative to prevent the depletion of resources that appears to be inevitable. Modern efforts at pollution control have flattened this curve slightly.

Graphic representation of Rome report.

This will pose problems in that industrialized nations are already feeling a loss in their standard of living and in non-industrialized nations that feel they have a right to achieve higher standards of living created by industrialization. The population growth continues upward and the supply of resources continues to diminish. With the increasing shortages of many minerals, we have been driven to search for new sources.

2.3.1 Environmental Effects of Extracting and using Mineral Resources

Most important environmental concern arises from the impacts of extraction and processing of the minerals during mining, melting, roasting, etc.

Mining

Mining is the process of extraction of metals from a mineral deposit.

Types of mining

The various types of mining are as follows:

Surface mining:

It is the process of extraction of raw materials from the near surface deposits.

Underground mining:

The process of extraction of raw materials below the earth's surface is known as underground mining. It includes:

i. Open-Pit mining

In this type of mining machines dig holes and remove the ores.

Example: Iron, Copper, Marble, etc.

ii. Dredging

In dredging, chained buckets and drag lines are used, which scrap up the minerals from wider-water mineral deposit.

iii. Strip Mining

In case of strip mining, the ore is stripped off by using bulldozers, stripping wheels.

Environmental Damage

The environmental damage, caused by mining activities is as follows:

Devegetation and defacing of landscape:

The top soil and the vegetation are removed from the mining area. Large scale deforestation or de vegetation leads to a number of ecological losses and also landscape gets badly affected.

Ground water contamination:

Mining disturbs and also pollutes the ground water. Usually sulfur, present as an impurity in many ores gets convert into disulphuric acid due to microbial action, which makes the water acidic. Some heavy metals also get leached into ground water.

Surface water pollution:

The drainage of acid from the mine often contaminates the nearby streams and lakes. The acidic water is harmful to many aquatic lives. Radioactive substances like Uranium also contaminate the surface water and kill many aquatic animals.

Air pollution:

Melting and roasting are done to purify the metals, which emits enormous amounts of air pollutants damaging the nearby vegetation. The suspended particulate matter, arsenic particles, cadmium, etc., contaminate the atmosphere and public suffer from several health problems.

Subsidence of land:

It is mainly associated with underground mining. Subsidence of mining area results in cracks in houses, tilting of buildings, bending of railway tracks.

Environment damages

- Mining activity not only destroys trees, it also pollutes soil, water and air with heavy metal toxins that are almost impossible to remove.

- Due to the continuous removal of minerals, forest covers, the trenches are formed on ground, leading to water logged area, which in turn contaminates the ground water.

- Destruction of natural habitat at the mine and waste disposal sites.

- During mining operations, the vibrations are developed, which leads to earthquakes.

- Noise pollution is another major problem from mining operations.

- When materials are disturbed in significant quantities during mining process, large quantities of sediments are transported by water erosion.

- Sometimes landslides may also occur as a result of continuous mining in forest area.

- Mining reduces the shape and size of the forest areas.

Environmental Impacts of Mining

The environmental responsibility of mining operations is the protection of the air, land and water. The mineral resources were developed in the United States for nearly two centuries with few environmental controls. This is largely attributed to the fact that environmental impact was not understood or appreciated as it is today. In addition, the technology available during this period was not always able to prevent or control environmental damage.

Air: All methods of mining affect air quality. Particulate matter is released in surface mining when overburden is stripped from site and stored or returned to the pit. When the soil is removed, vegetation is also removed, exposing the soil to the weather, causing particulates to become airborne through wind erosion and road traffic. Particulate matter can be composed of such noxious materials as arsenic, cadmium and lead. In general, particulates affect human health adversely by contributing to illnesses relating to the respiratory tract, such as emphysema, but they also can be ingested or absorbed into the skin.

Land (Sand): Mining can cause physical disturbances to the landscape, creating eyesores such as waste rock piles and open pits. Such disturbances may contribute to the decline of wildlife and plant species in an area. In addition, it is possible that many of the pre-mining surface features cannot be replaced after mining ceases. Mine subsidence can cause damage to buildings and roads.

Water: Water-pollution problems caused by mining include acid mine drainage, metal contamination and increased sediment levels in streams. Sources can include active or abandoned surface and underground mines, processing plants, waste-disposal areas, haulage roads or tailings ponds. The sediments, typically from increased soil erosion, cause siltation or the smothering of streambeds. This siltation affects fisheries, swimming, domestic water supply, irrigation and other uses of streams.

Acid mine drainage (AMD) is a potentially severe pollution hazard that can contaminate surrounding soil, groundwater and surface water. The formation of acid mine drainage is a function of the geology, hydrology and mining technology employed at a mine site. The primary sources for the acid generation are sulfide minerals, such as pyrite (iron sulfide), which decompose in air and water. Many of these sulfide minerals originate from waste rock removed from the mine or from tailings. If water infiltrates pyrite-laden rock in the presence of air, it can become acidified, often at a pH level of two or three. This increased acidity in the water can destroy living organisms and corrode culverts, piers, boat hulls, pumps and other metal equipment in contact with the acid waters and render the water unacceptable for drinking or recreational use.

2.3.2 Sustainable Mining of Granite, Literate, Coal, Sea and River Sands

We define sustainability in mining as mining and mineral development that meets the growing needs of all communities while maintaining a healthy environment and vibrant economy for present and future generations.

The Criteria in Sustainability in Mining can be applied to mineral exploration and mining projects/operations as follows:

- Health and Safety: The project/operation is acting to ensure the health and safety of workers and the community.

- Effective Engagement: The relationships with those affected by a project/operation are characterized by integrity and trust.

- Respect for Indigenous Peoples: The project/operation respects the rights, culture and values of the Indigenous Peoples.

- Environment: Actions are being taken to ensure the maintenance and strengthening of environmental integrity over long term within the region of influence of the project/operation.

- Full mine or Operation Life Cycle: A full mine or operation life cycle perspective is being applied for planning and decision making that spans exploration through post-closure.

- Resource-use Efficiency: The project/operation is seeking to minimize resource inputs-energy, water, reagents, supplies, etc. while also minimizing contaminant outputs to air, water and land.

- Continuous Learning and Adaptation: The uncertainty inherent in mining operations is recognized and a commitment to continuous learning is displayed.

- Benefits: The project/operation is enhancing the potential for creating economic, social and cultural benefits for the local community or region.

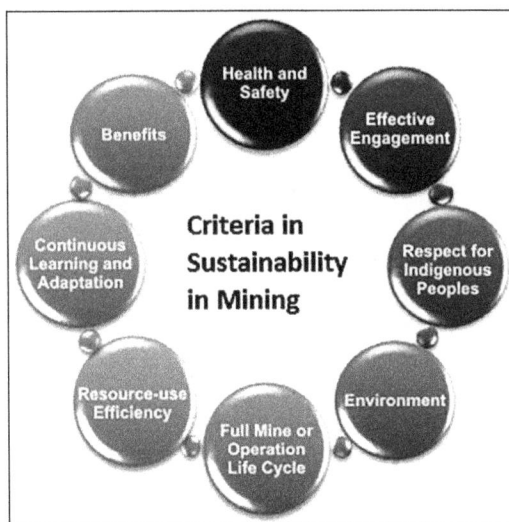

The importance of sustainable development principles has been increasing within the mining sector over the past two decades. Early work focused mainly on mining metals and commodities other than coal and energy fuels. Because sustainability, although, is an important consideration for all human endeavors now, the coal industry has become active in sustainability efforts. A number of global coal mining companies have embraced sustainability as a key aspect of corporate philosophy.

Continued production of the minerals and fossil energy fuels may not fit into commonly understood definitions of sustainability. The mineral and energy extraction and recla-mation operations do, however, contribute significantly to sustainability through the benefits they provide to society, when they are conducted in a manner that supports sustainable economies, social structures and environments throughout all phases of mining, including closure.

Significant progress can also be made through the inclusion of sustainability concepts in the original design of the operation, as well as in ongoing operations. Innovative en-gineering, mining and reclamation operations can be optimized through consideration of environmental and economic sustainability goals, side-by-side with traditional tech-nical mining engineering considerations.

It is widely recognized that coal is and will continue to be a crucial element in a modern, balanced energy portfolio, providing a bridge to the future as an important low cost and secure energy solution to sustainability challenges. In response, global coal and energy production industries have begun a major effort to identify and accelerate the deployment and further development of innovative, advanced, efficient, cleaner coal technologies. A number of coal producers are also involved in sustainable development activities, including economic support of communities and regions, environmental res-toration and social well-being.

The designer of, specially, coal mining operations needs to simultaneously consider legal, environmental and sustainability goals along with traditional mining engineering parameters as an integral part of the design process. The role of coal in global energy supply mix makes this of primary importance. There is a need for research into the parameters for mining design that allow the building of models for optimization, the relationships between those parameters and desired outcomes that the system is being optimized to produce. In addition to quantifying the economic viability of the opera-tion, a number of sustainability goals should be built into the model and the relative importance of those goals determined.

2.4   Food Resources: World Food Problems

Food is an essential requirement for human survival. Each person has minimum food requirement. The main components of food are carbohydrates, fats, proteins, minerals and vitamins.

Types of Food Supply

Historically humans have a dependency on three systems for their food supply:

  • Croplands: It mostly produces the grains and provides around 76% of the world's food. Example: Rice, Wheat, Maize.

- Range lands: It produces food mainly from grazing livestock and provides around 17% of the world's food. Example: Meat, Milk, Fruits.

- Oceans: Oceanic fisheries supply about 7% of the world's food. Example: Fish, Prawn, Crab, etc.

Food Problems

We know that 79% of the total area of the earth is covered with water. Only 21% of the earth surface is land, of which most of the areas are desert, forest, mountains and barren areas. Only a minimal percentage of land is for cultivation, which in turn often does not suffice for the ever growing world population, as day by day the world population increases, while the cultivable land area decreases.

The environmental degradation like soil erosion, water logging, water pollution, salinity and affects agricultural lands. Urbanization is another problem in developing countries, which deteriorates the agricultural lands.

Since the food grains like wheat, rice, corn and vegetable like potato are the major food for the people all over the world, the food problem raises.

A key problem is the human activity which degrades most of the earth is net primary productivity which supports all life.

We know that 79% of total area of the earth is covered with water. Only 21% of the earth surface is land, of which most of the areas are desert, forest, mountains and barren areas. Only a minimal percentage of land is for cultivation, which in turn often does not suffice for the ever growing world population, as day by day the world population increases, while the cultivable land area decreases.

Environmental degradation like soil erosion, water logging, water pollution, salinity, affect agricultural lands.

Urbanization is another problem in developing countries, which deteriorates the agricultural lands.

Since the food grains like rice, wheat, corn and the vegetable like potato are the major food for the people all over the world, the food problem raises.

A key problem is the human activity which degrades most of the earth is net primary productivity which supports all life.

Under Nutrition and Malnutrition

Nutrition or Nutritious or Nourished:

To maintain good health and resist disease, we need large amount of the macro nutrients such as carbohydrates, proteins and fats and smaller amount of macro nutrients such as Vitamin A, C and E and minerals such as calcium, iron and iodine.

The food and agriculture organization (FAO) of United Nation estimated that on an average, the minimum caloric intake on a global scale is 2,500 calories/day.

Under Nutrition or Under Nourished:

People who cannot afford enough food to meet the basic energy needs suffer from under nutrition. They receive less than 90% of the minimum dietary calories.

Effect of under nutrition: Suffer from mental retardation and infectious diseases such as and diarrhea.

Malnutrition or Malnourished:

Besides the minimum caloric intake we also need proteins, minerals, vitamins, iron and iodide. Deficiency of lack of nutrition often leads to malnutrition resulting in several diseases.

Effect of Malnutrition

S. No.	Deficiency of Nutrient	Effects
1.	Proteins	Growth
2.	Iron	Anemia
3.	Iodide	Goiter, Cretinism
4.	Vitamin A	Blindness

Thus, chronically under nourished and malnourished people are diseases prone and are too weak to work or think clearly.

World Food Problems

- Environmental degradation like soil erosion, water pollution, water logging, salinity, affect agricultural lands.

- Urbanization is another problem in developing countries, which deteriorates agricultural lands.

- A key protein is the human activity, which degrade most of the world's net primary productivity which supports all life.

2.4.1 Changes Caused by Non-agriculture Activities

Non-Agriculture Activities

The different types of non-farm activities can be performed by farmers in rural areas.

These activities and their dominance vary from place to place. In the study sites, the predominant non-farm activities are trading and handicraft.

Trade is predominant in Kachabira Wereda, while handicraft, mainly weaving, is the predominant activity in Damotgale Wereda. Among the sample population, there are only six households who reported non-farm activity other than trading and handicraft in the study sites.

There are three households reported sale of `Injera' (traditional staple food) and bread; two farmers are employed as labourers while one farmer earns income from renting cattle.

Prior to 1974, population in Kambata and Welaita were known for their seasonal migration to plantation sites and commercial farms. The sample population did not indicate any of this pattern at present.

In a similar way, a study by Getachew (1995) also found the insignificance of labour migration at present. About 162 farmers or 70.1 percent reported that they earn income from non-farm activities. The remaining 67 farmers or 29.2 percent of our sample have no income from non-farm activity the mean income from non-farm activities is 633.95 birr.

The highest income is recorded for Lesho and Gemesha in Kachabira Weredas, where 1255.33 and 578.37 birr, respectively, were reported. Distribution of income from non-farm activities shows that about 46.2 percent earn in the range of 1-500 birr.

The drops to 16.2 percent for income group between 500-1000 birr. Those farmers earning above 1500 birr are only 5.7 percent of the total.

Changes Caused by Non-agriculture Activities

An engagement in nonagricultural activities in rural areas can be classified into survival-led or opportunity-led. Survival-led diversification would decrease inequality by increasing the incomes of poorer households and thus reduce poverty.

By an opportunity-led diversification would increase inequality and has a minor effect on poverty, as it tends to be confined to nonpoor households. Using data from Western Kenya, we confirm the existence of the differently motivated diversification strategies. Yet, the poverty and inequality implications differ somewhat from the expectations.

Our findings indicate that in addition to asset constraints, rural households also face limited or relatively risky high return opportunities outside the agriculture.

2.4.2 Effects of Modern Agriculture

- Micro nutrient imbalance: Most of the chemical fertilizers used in modern agriculture contain nitrogen phosphorus and potassium (N.R.K) which are macro nutrients when excess of the fertilizers are used in the fields, it causes micro nutrient imbalance.

- Blue Baby syndrome (Nitrate Pollution): When the Nitrogenous fertilizers are applied in the they leach deep into the soil and contaminate the ground water. The nitrate concentration in the water gets increased. When the concentration exceeds 25 mg 1 lit, they cause serious health problem called Blue Baby syndrome. This disease mainly affects infants and can even lead to death.

- Eutrophication: A large proportion of N and P fertilizers used in crop fields is washed off by the runoff water and reaches the water bodies causing over nourishmnet of the lakes. This process is called as Eutrophication.

Due to eutrophication the lakes get attacked by algal blooms. These species use up the nutrients rapidly and grow very fast. Since the life time of the global species are less they die quickly and pollute the water, which in turn affect the aquatic life.

2.4.3 Fertilizer-pesticide Problems

Problems of Pesticides on Modern Agriculture

In order to improve the crop yield, lots of pesticides are used in the agriculture.

First generation pesticides:

Sulfur, arsenic, lead or mercury is used to kill the pests.

Second generation pesticides:

DDT is used to kill the pests. Although these pesticides protect our crops from huge losses due to pests, they produce number of side effects such as:

- Death of non-target organisms.

- Producing new pests.

- Bio magnification.

- Risk of cancer.

Effect of Pesticides:

- It directly acts as Carcinogens.

- It indirectly suppresses the immune system, the soil and contaminant the ground water. The nitrate concentration in the coater gets increased. When the nitrate concentration exceeds 25 mg/lit., they cause serious health problem called, "Blue Baby Syndrome". This disease affects infants and leads even to death.

Derived qualities of an ideal pesticide:

- An ideal pesticide must laid only the target species.

- It must be a biodegradable.

- It should not produce new pests.

- It should not produce any toxic pesticide vapor.

- Excessive synthetic pesticide should not be used.

- Chlorinated pesticides and phosphate pesticides are hazardous, so they should not be used.

2.4.4 Water Logging

Water logging refers to saturation of soil with water.

Problems in Water Logging

During water logged conditions, the soil gets filled with water and the soil air gets depleted. In such conditions the roots of the plants do not get enough air for respiration. So, mechanical strength of the soil decreases and crop yield fairs.

Causes of water logging

- Heavy rain.

- Excessive water supply to the crop lands.

- Poor drainage.

Remedy

Preventing excessive irrigation, sub surface drainage technology and bio drainage by trees like eucalyptus tree are some method of preventing water logging.

2.4.5 Salinity

The water, not absorbed by the soil, undergoes evaporation's leaving behind a thin layer of dissolved salts in the topsoil. The process of accumulation of salts is called salinity of the soil. The salts are characterized by the accumulation of soluble salts like calcium chloride, sodium chloride, sodium sulfate, magnesium chloride, sodium bicarbonates and sodium carbonates. The pH of the water exceeds 8.0.

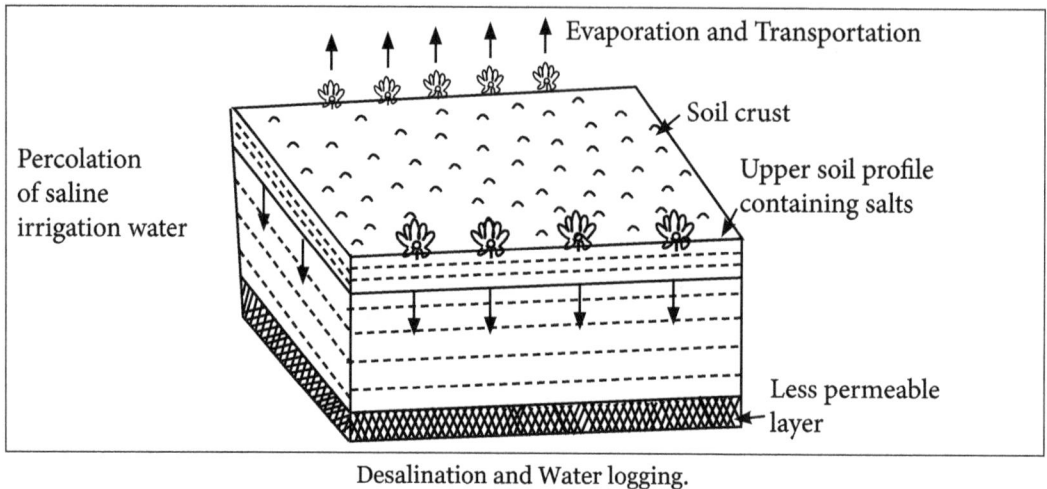

Desalination and Water logging.

Problems in Salinity

Most of the water, used for irrigation comes only from canal or ground, which unlike rainwater contains dissolved salts. Under dry climate or the water gets evaporated leaving behind the salt in the upper portion of the soil. Due to salinity the soil becomes alkaline and crop yield decreases.

Remedy:

- The salt deposit is removed by flushing them out by applying higher grade quality water to such soils.

- Using sub surface drainage system the salt water is flushed out slowly.

2.5   Energy Resources: Growing Energy Needs

Sources of Energy

Energy may be defined as, "any property, which can be converted into work".

Energy is available on the earth in a number of forms, some of which may be immediately used to do work, while the others require some process of transformation. All the development activities in world are directly or indirectly dependent upon energy. Both energy production and energy utilization are the indicators of a country's progress.

Development of Energy

First form of energy is the fire. The early mass discovered fire and used it for cooling and heating purposes. Wood is the main source of energy, which is later replaced by coal. Coal is new being replaced by oil and gas. Now due to insufficient availability and price hike, people started of thinking and using several alternate sources of the energy.

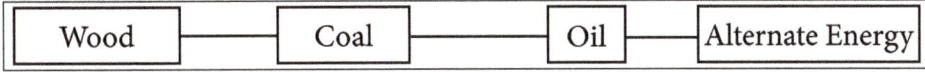

Wood	Coal	Oil	Alternate Energy

Sourcing Energy Needs

Energy is essential to all human societies. All industrial process like lighting, mining, transport, heating and cooling in buildings, all require energy. With the demands of the growing population, the world is facing further energy deficit.

Our life style is also changing from sample way of life to luxurious life style. At present 95% of the commercial energy is available only from the fossils fuels like oil, coal and natural gas are not going to last for many more years. It would be really ironic if fuel becomes more expensive than the food.

Energy Distribution World Scenario

Developed countries like U.S.A and Canada constitute only 5% of the world's population, but consume 25% of the available world's energy resources. It has been observed, that in USA and Canada an average person consumer 300GJ per year.

So a person in a developed country consumes almost as much energy in a single day as one person consumes in a whole year in a poor country. From the scenario above, it is clear that own life style and standard of living are closely related to energy needs.

Per Capita Energy Use

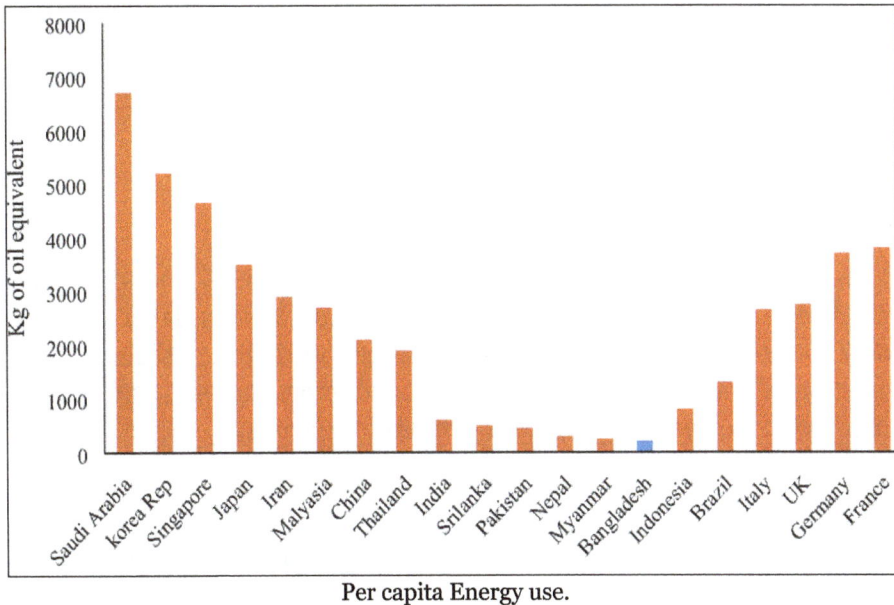

Per capita Energy use.

The correlation is obtained between per capita energy use and GNP. The developed countries like U.S.A., Japan, Switzerland, etc., with help GNP show high energy use while poor countries like India, China, Ethiopia with GNP show low energy use.

2.5.1 Renewable Energy Resources

Solar Energy

The energy that we get directly from the sun is called solar energy.

The nuclear fusion reactions occurring inside the sun release enormous amount of energy in the form of heat and light. Various techniques are available for collecting, converting and using the solar energy.

Methods of Harvesting Solar Energy

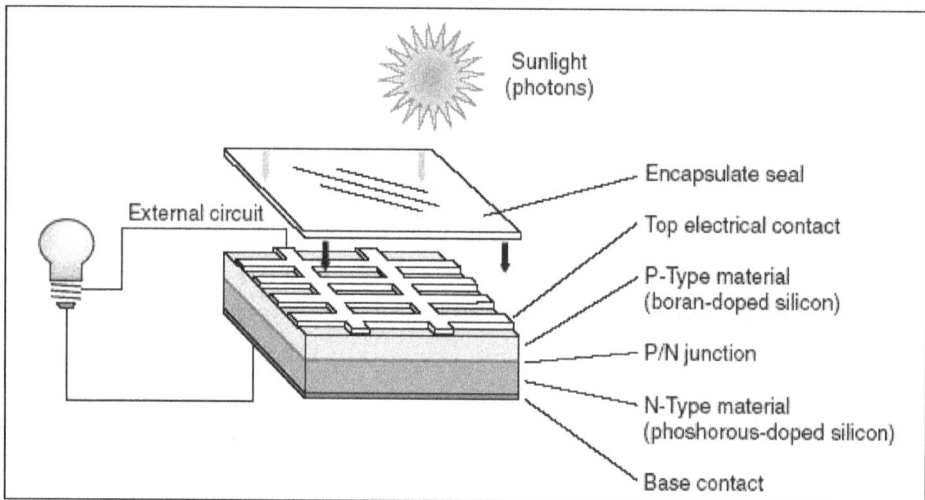

Solar cell.

Some important solar energy, harvesting devices are given below:

Solar cells or Photovoltaic Cells or PV cells:

Solar cells consist of a p-type semiconductor and n-type semiconductor. They are in close contact with each other. When solar rays fall on the top layer of p-type semiconductor. The electrons from the valence band get promoted to the conduction band and cross the p-n junctions into n-type semiconductor, thereby potential difference between two layers is created which causes flow of electrons.

Uses:

Solar cells are used in electronic watches, calculators, water pumps, street lights and to run radio and TVs.

Solar Battery:

When a large number of solar cells are connected in series it forms a solar battery. Solar battery produces electricity capable of running water pump to run street-light etc. They are used in remote areas where conventional electricity supply is a problem.

Solar Water Heater:

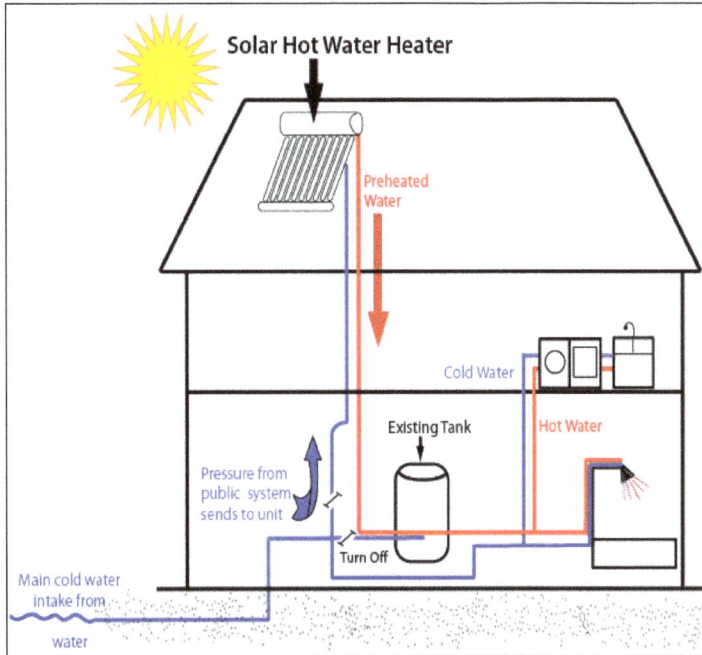

Solar water heater.

It consists of an insulated box inside of which is painted with black paint. It is also provided with a glass lid to receive and store the solar heat. Inside the box, it has a black painted copper coil, through which cold water is allowed to flow in, which gets heated up and flows out into a storage tank. From the storage tank water is then supplied through pipes.

Significance of solar energy:

- Solar water heaters, cookers, require neither fuel not attention while cooking the food.

- Solar cells are free from air and sound pollution.

- Solar cells can be used in remote and isolated areas, forests, hilly regions.

Solar Heat Collectors:

Solar heat collectors consists of the natural materials like stones, bricks or materials like glass, which can absorb heat during the day time and release it slowly at night.

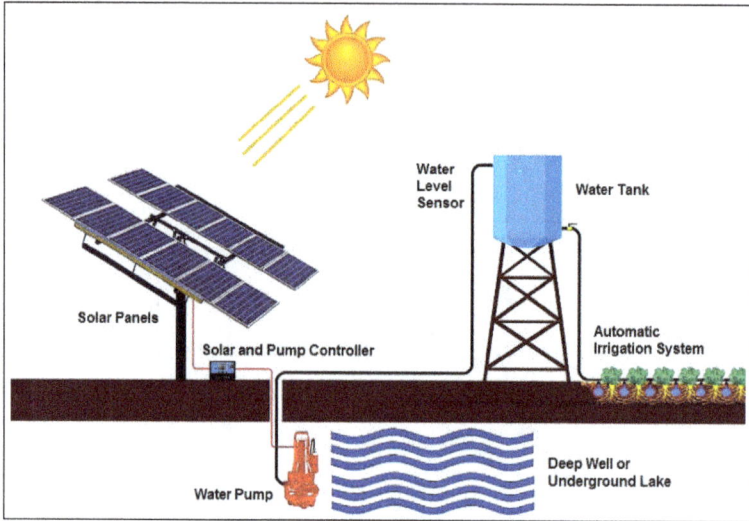

Solar pump run by solar cells.

Uses:

It is generally used in cold places, where the houses are kept in hot condition using solar heat collectors.

Wind Energy

Moving air is called wind. Energy covered from the force of the wind is called wind energy. The energy possessed by wind is because of its high speed. The wind energy is harnessed by making use of wind mills.

Wind Mills:

The strike on the blades of the wind will make it rotating continuously. The rotational motion of blades drives a number of machines like water pump, flour mills and electric generators.

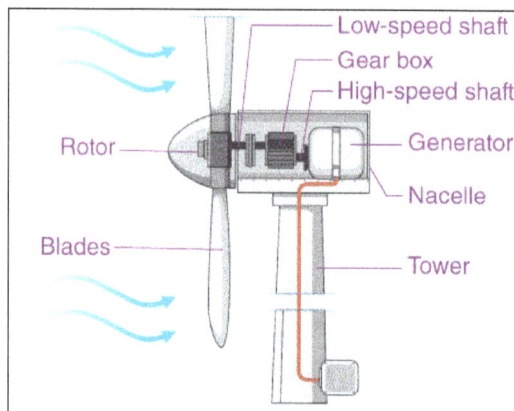

Wind mill.

Wind Forms: When a large number of wind mills are installed and joined together in definite pattern it forms a wind farm. The wind farms produce large amount of electricity.

Condition: The minimum speed required for satisfactory working of a wind generator is 15 km/hr.

Advantages:

- It is very cheap.

- It does not cause any air pollution.

Significance of Wind Energy:

- It is made available easily in many off-shore, on-shore and remote areas.

- The generation period of wind energy is low and power generation starts from commissioning.

- It is recommended to broaden the nation's energy options for new energy sources.

Ocean Energy

Ocean can also be used for generating energy in the following ways:

Ocean Thermal Energy: There is often a large temperature difference between the surface level and deeper level of the tropical oceans. This temperature difference can be utilized to generate electricity. The energy available due to the difference in temperature of water is called thermal energy.

Condition: The temperature difference should be 20°C or more is required between surface water and ground water.

Process: The warm surface water of ocean is used to boil a low boiling liquid like ammonia. The high vapor pressure of the generator and generates electricity. The cold water from the deeper ocean is pumped to cool and condense the vapor into liquid.

Significance of OTE:

- OTE is continuous, renewable and pollution free.

- The use of cold deep water as the chiller fluid in air-conditioning has also been proposed.

- Electric power generated by OTE can be used to produce hydrogen.

Geothermal Energy

The temperature of the earth increases at a rate of 20 - 75°C per km, when we move down the earth surface. High temperature and high pressure steam fields exists below the earth's surface in many places. The energy harnessed from the high temperature present inside the earth is called geothermal energy.

Natural geysers:

In some of the places, hot water or steam comes out of the ground through cracks naturally in the form of natural geysers.

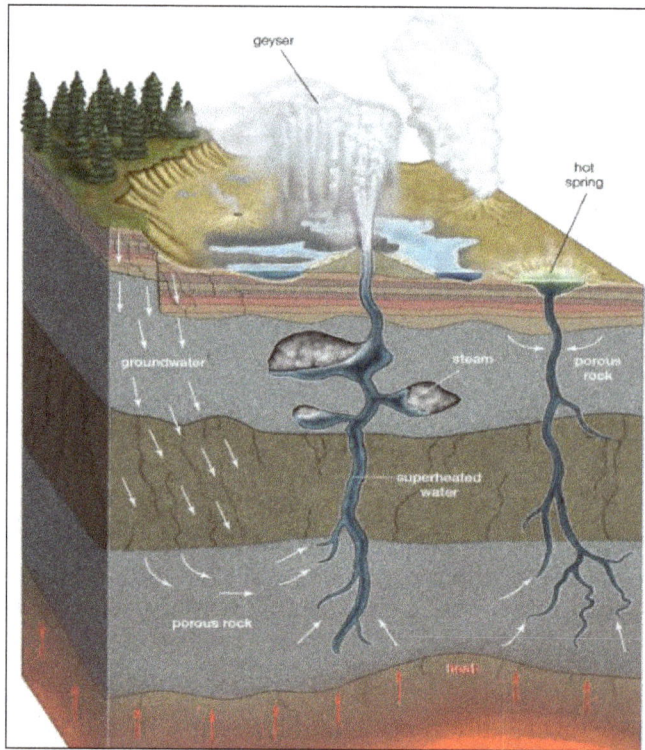

Artificial geysers.

Artificial geysers:

In some of the places, we can artificially dry a hole up to the hot region and by sending a pipe in it we can make hot water or steam to rush out through the pipe with very high pressure.

Thus, the hot water or steam coming out from the natural or artificial geysers is allowed to rotate the turbine of a generator to produce electricity.

Significance of Geo-Thermal Energy:

*   The power generation level is higher for geothermal than for solar and wind energies.

- Geothermal power plants can be brought on line more quickly than most other energy sources.

- GTE is effectively and efficiently used for direct uses such as hot water bath, resorts and aquaculture green houses.

Tidal Energy or Tidal Power

Ocean tides, produced by gravitational forces of sun and moon, contain enormous amount of energy. The "high tide" and "low tide" refer can be to the rise and fall of water in the oceans. The tidal energy shall be harnessed by constructing a tidal barrage.

During high tide, the seawater is allowed to flow into the reservoir of barrage and rotates the turbine which in turn produces electricity by rotating the generators.

During low tide, when the sea level is low, the sea water stored in the barrage reservoir is allowed to flow in and in turn rotating the turbine again.

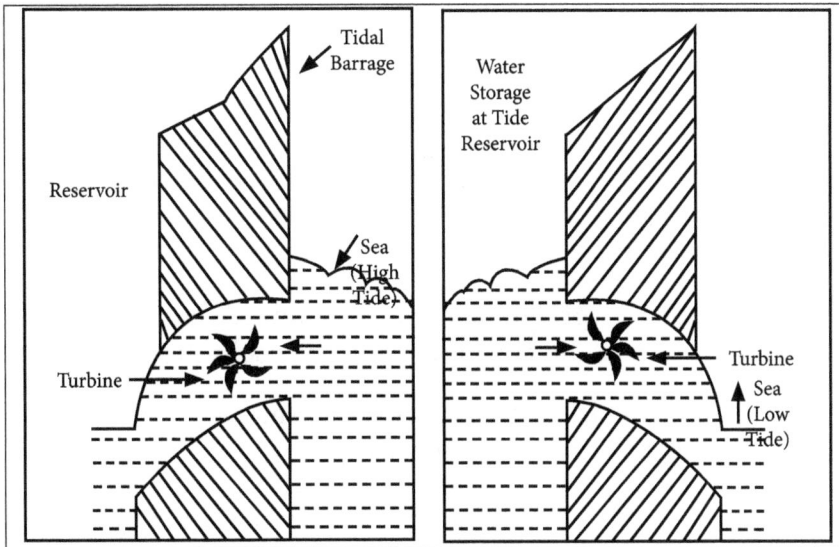

(a) Water flows into the reservoir from sea (b) Water flows out from the reservoir to the sea.

Significance of Tidal Energy:

- Tidal power plants do not require large areas of valuable lands as they are on the bays or estuaries.

- As the sea water is inexhaustible, it is completely independent of the uncertainty of precipitations.

- It is pollution free energy source as it does not use any fuel and also does not produce any wastes.

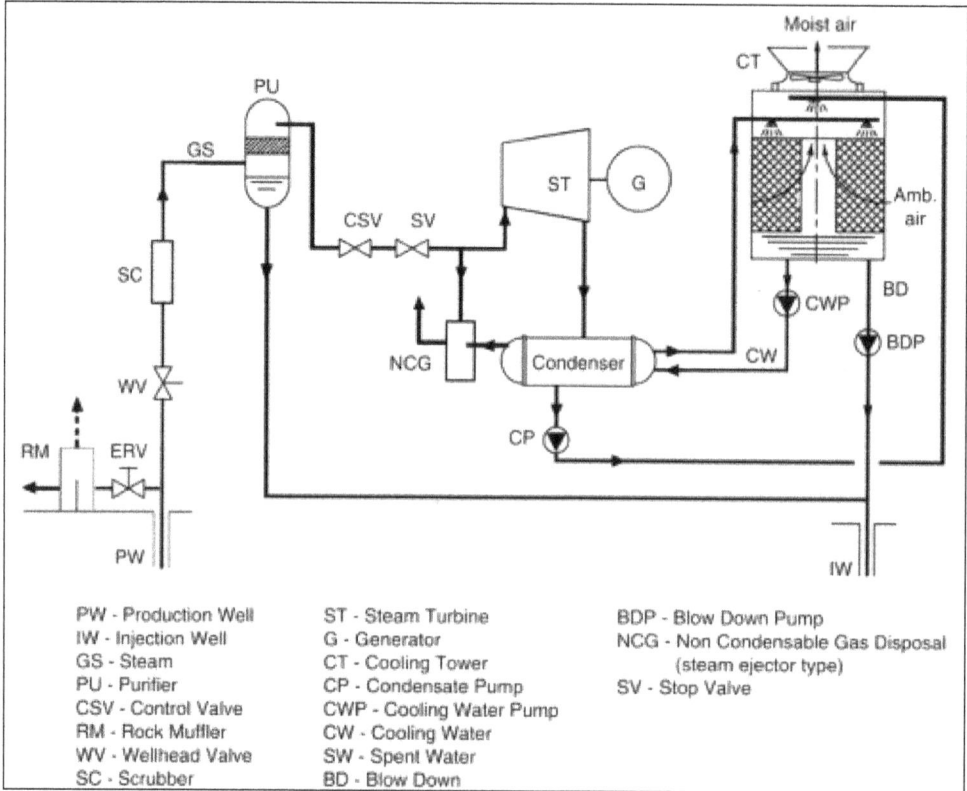

PW - Production Well	ST - Steam Turbine	BDP - Blow Down Pump
IW - Injection Well	G - Generator	NCG - Non Condensable Gas Disposal
GS - Steam	CT - Cooling Tower	(steam ejector type)
PU - Purifier	CP - Condensate Pump	SV - Stop Valve
CSV - Control Valve	CWP - Cooling Water Pump	
RM - Rock Muffler	CW - Cooling Water	
WV - Wellhead Valve	SW - Spent Water	
SC - Scrubber	BD - Blow Down	

Geothermal power plants.

Biomass Energy

Biomass is the organic matter, produced by plants or animals used as sources of energy. Most of the biomass is burned directly for heating and cooling and industrial purpose.

Examples: Wood, crop residues, seeds, cattle dung, sewage, agricultural wastes, etc. Biomass energies are of any one of the following types:

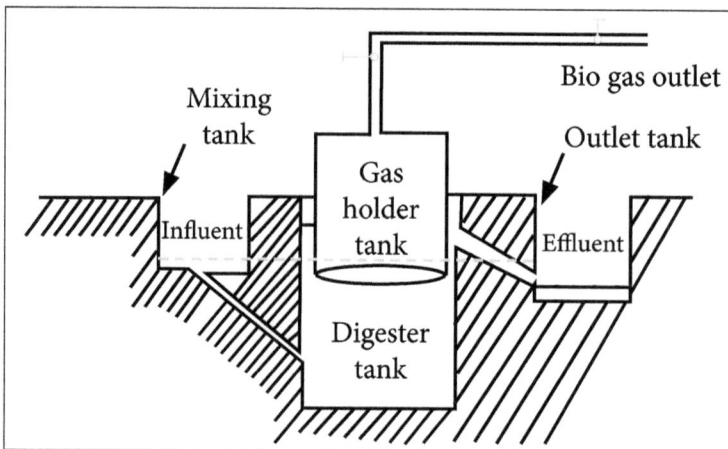

Biogas plant.

Bio gas:

Bio gas is a mixture of gases such as methane, carbon dioxide, hydrogen sulfide, etc. It contains about 65% of methane gas as a major constituent.

Bio gas is obtained by the anaerobic fermentation of animal dung or plant wastes in the presence of water.

Bio fuels:

Bio fuels are the fuels, obtained by the fermentation of bio mass.

Examples: Ethanol, Methanol.

- Ethanol: Ethanol can be produced from sugarcane. Its calorific value is less when compared to petrol and produces much less heat than petrol.

- Methanol: Methanol can be obtained from ethanol or sugar containing plants. Its calorific value is also too low when compared to diesel and gasoline.

- Gasohol: Gasohol is a mixture of ethanol and gasoline. In India, a trial is being carried out to use Gasohol in cars and buses.

Hydrogen fuel:

Hydrogen can be produced by thermal dissociation or photolysis or electrolysis of water. It possess high calorific value. It is nonpolluting, because the combustion predict is water.

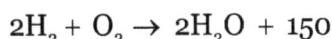

$$2H_2 + O_2 \rightarrow 2H_2O + 150$$

Disadvantages of hydrogen fuel:

- Safe handling is required.

- Hydrogen is difficult to store and transport.

- Hydrogen is highly inflammable and explosive in nature.

Nonrenewable Energy Resources

Coal

Coal is a solid fossil full formed in several stages as buried remains of land plants that lived 300–400 million years ago were subjected to intense heat and pressure over millions of years.

The various stages of coal during the coalification of wood are:

Wood → Peat → Lignite → Bituminous

The carbonate content of anthracitic is 90% and its calorific value is 8700 k.cal. The carbon content of bituminous, lignite, pert and wood are 80, 70, 60 and 50% respectively.

Disadvantages:

- When coal is burnt it produces CO_2, which causes global warming.

- Since coal contains impurities like S and N, it produces toxic gases during burning.

i. Natural Gas

Natural gas is found above the oil in oil well. It is a mixture of 50-90% methane and a small amount of other hydrocarbons. Its calorific value ranges from 12,000- 14,000 k.cal/m^3.

ii. Dry Gas

If the natural gas contains lower hydrocarbons like methane and ethane, it is called dry gas.

iii. Wet Gas

If the natural gas contains higher hydrocarbons like propane, butane along with methane it is called wet gas.

Like petroleum oil, natural gas can also be formed by the decomposition of dead animals and plant that were buried under lake and ocean, at high temperature and pressure for millions of years.

Crude Oil

It is a mixture of hydrocarbons that formed from plants and animals that lived millions of years ago. It is a fossil fuel and it exists in liquid form in underground pools or reservoirs, in tiny spaces within sedimentary rocks and near the surface in tar sands. Petroleum products are fuels made from crude oil and other hydrocarbons contained in natural gas. Petroleum products can also be made from coal, natural gas and biomass.

Products made from Crude Oil

After crude oil is removed from the ground, it is sent to a refinery where different parts of the crude oil are separated into usable petroleum products. These petroleum products include gasoline, distillates such as the diesel fuel and heating oil, jet fuel, petrochemical feedstock's, waxes, lubricating oils and asphalt.

A 42 U.S. gallon barrel of crude oil yields about 45 gallons of petroleum products in U.S. refineries because of refinery processing gain. This increase in volume is similar to what happens to popcorn when it is popped.

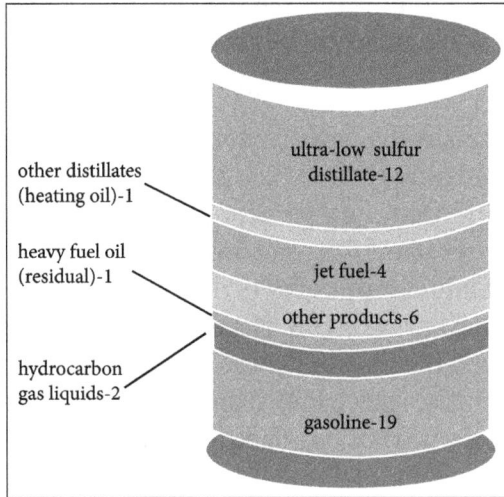

2.5.3 Use of Alternate Energy Sources: Oil and Natural Gas Extraction

Renewable energy wind, solar, geothermal, hydroelectric and biomass provides substantial benefits for our climate, our health and our economy. Each source of renewable energy has unique benefits and costs.

Little to no Global Warming Emissions

The human activity is overloading our atmosphere with carbon dioxide and other global warming emissions, which trap heat, steadily drive up the planet's temperature and create significant and harmful impacts on our health, environment and climate.

Electricity production accounts for more than one-third of U.S. global warming emissions, with the majority generated by coal-fired power plants, which produce approximately 25 percent of total U.S. global warming emissions, natural gas-fired power plants produce 6 percent of total emissions. In contrast, most renewable energy sources produce little to no global warming emissions.

According to data aggregated by International Panel on Climate Change, life-cycle global warming emissions associated with renewable energy including manufacturing, installation, operation and maintenance and dismantling and decommissioning are minimal.

Compared with natural gas, which emits between 0.6 and 2 pounds of carbon dioxide equivalent per kilowatt-hour (CO_2E/kWh) and coal, which emits between 1.4 and 3.6

pounds of CO_2E/kWh, wind emits only 0.02 to 0.04 pounds of CO_2E/kWh, solar 0.07 to 0.2, geothermal 0.1 to 0.2 and hydroelectric between 0.1 and 0.5. Renewable electricity generation from biomass can have a wide range of global warming emissions depending on the resource and how it is harvested. Sustainably sourced biomass has a low emissions footprint, while unsustainable sources of biomass can generate significant global warming emissions.

Increasing the supply of renewable energy would allow us to replace carbon-intensive energy sources and significantly reduce U.S. global warming emissions. For example, a 2009 UCS analysis found that a 25 percent by 2025 national renewable electricity standard would lower power plant CO_2 emissions 277 million metric tons annually by 2025—the equivalent of annual output from 70 typical (600 MW) new coal plants. In addition, a ground-breaking study by U.S. Department of Energy's National Renewable Energy Laboratory explored the feasibility and environmental impacts associated with generating 80% of the country's electricity from renewable sources by 2050 and found that global warming emissions from the electricity production could be reduced by approximately 81 percent.

Improved Public Health and Environmental Quality

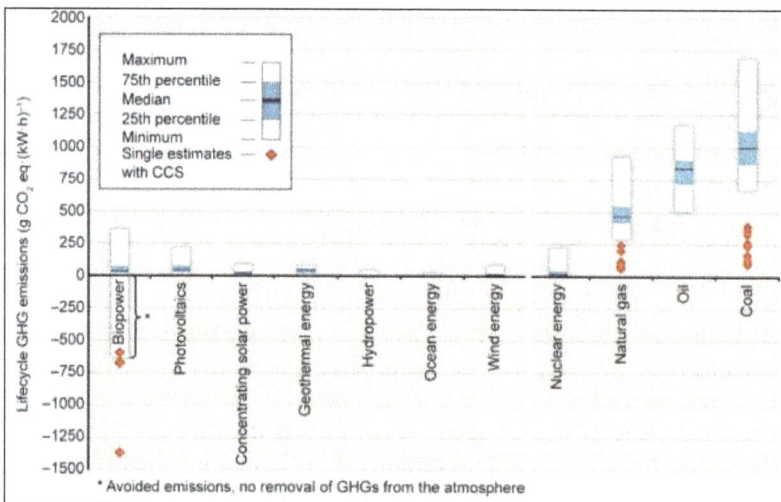

Generating electricity from renewable energy rather than fossil fuels offers significant public health benefits. The air and water pollution emitted by coal and natural gas plants is linked to breathing problems, neurological damage, heart attacks and cancer. Replacing fossil fuels with renewable energy has been found to reduce premature mortality and lost workdays and it reduces overall healthcare costs. The aggregate national economic impact associated with these health impacts of the fossil fuels is between $361.7 and $886.5 billion or between 2.5% and 6% of gross domestic product (GDP).

Wind, solar and hydroelectric systems generate electricity with no associated air pollution emissions. While geothermal and biomass energy systems emit some air pollutants, total air emissions are normally much lower than those of coal- and natural gas-fired power plants.

In addition, wind and solar energy require essentially no water to operate and thus do not pollute water resources or strain supply by competing with agriculture, drinking water systems or other important water needs. In contrast, fossil fuels can have a significant impact on water resources. For example, both coal mining and natural gas drilling can pollute sources of drinking water. Natural gas extraction by hydraulic fracturing (fracking) requires large amounts of water and all thermal power plants, including those powered by the coal, gas and oil, withdraw and consume water for cooling.

Biomass and geothermal power plants, like coal- and natural gas-fired power plants, require water for cooling. In addition, hydroelectric power plants impact river ecosystems both upstream and downstream from the dam. However, NREL's 80 percent by 2050 renewable energy study, which included biomass and geothermal, found that water withdrawals would decrease 51 percent to 58 percent by 2050 and water consumption would be reduced by 47 to 55%.

A Vast and Inexhaustible Energy Supply

Throughout the United States, strong winds, sunny skies, plant residues, heat from the earth and fast-moving water can each provide a vast and constantly replenished energy resource supply. These diverse sources of renewable energy have the technical potential to provide all electricity the nation needs many times over.

Estimates of the technical potential of each renewable energy source are based on their overall availability given certain technological and environmental constraints. In 2012, NREL found that together, renewable energy sources have the technical potential to supply 482,247 billion kilowatt-hours of electricity annually. This amount is 118 times the amount of electricity the nation currently consumes. Although, it is important to note that not all of this technical potential can be tapped due to conflicting land use needs, the higher short-term costs of those resources, constraints on ramping up their use such as limits on transmission capacity, barriers to public acceptance and other hurdles.

Today, renewable energy provides only a tiny fraction of its potential electricity output in the United States and worldwide. But numerous studies have repeatedly shown that renewable energy can be rapidly deployed to provide a significant share of future electricity needs, even after accounting for potential constraints.

Jobs and other Economic Benefits

Compared with the fossil fuel technologies, which are typically mechanized and capital intensive, the renewable energy industry is more labor-intensive. This means that, on average, more jobs are created for each unit of electricity generated from the renewable sources than from fossil fuels.

Renewable energy already supports thousands of jobs in United States. For example, in 2011, the wind energy industry directly employed 75,000 full-time-equivalent employees in a variety of capacities, including manufacturing, project development, construction and turbine installation, operations and maintenance, transportation and logistics, financial, legal and consulting services. More than 500 factories in the United States manufacture parts for wind turbines and the amount of domestically manufactured equipment used in wind turbines has grown dramatically in recent years: from 35 percent in 2006 to 70 percent in 2011.

Other renewable energy technologies employ even more workers. In 2011, the solar industry employed approximately 100,000 people on a part-time or full-time basis, including jobs in solar installation, manufacturing and sales, the hydroelectric power industry employed approximately 250,000 people in 2009 and in 2010 the geothermal industry employed 5,200 people.

Increasing renewable energy has the potential to create still more jobs. In 2009, the Union of Concerned Scientists conducted an analysis of economic benefits of a 25 percent renewable energy standard by 2025, it found that such a policy would create more than three times as many jobs as producing an equivalent amount of electricity from the fossil fuels resulting in a benefit of 202,000 new jobs in 2025.

In addition to the jobs directly created in the renewable energy industry, growth in renewable energy industry creates positive economic "ripple" effects. For example, industries in renewable energy supply chain will benefit and unrelated local businesses will benefit from increased household and business incomes.

In addition to creating new jobs, increasing our use of renewable energy offers other important economic development benefits. Local governments collect property and income taxes and other payments from the renewable energy project owners. These revenues can help support vital public services, especially in rural communities where projects are often located. Owners of the land on which wind projects are built also often receive lease payments ranging from $3,000 to $6,000 per megawatt of installed capacity, as well as payments for power line easements and road rights-of-way or they may earn royalties based on the project's annual revenues. Similarly, farmers and rural landowners can generate new sources of supplemental income by producing feedstocks for biomass power facilities.

UCS analysis found that a 25 by 2025 national renewable electricity standard would stimulate $263.4 billion in new capital investment for the renewable energy technologies, $13.5 billion in new landowner income biomass production and/or wind land lease payments and $11.5 billion in new property tax revenue for local communities.

Renewable energy projects therefore keep money circulating within the local economy and in most states renewable electricity production would reduce the need to spend money on importing the coal and natural gas from other places. Thirty-eight states were net importers of the coal in 2008 from other states and, increasingly, other countries: 16 states spent a total of more than $1.8 billion on coal from as far away as Colombia, Venezuela and Indonesia and 11 states spent more than $1 billion each on net coal imports.

Stable Energy Prices

The renewable energy is providing affordable electricity across the country right now and can help stabilize energy prices in future.

The costs of renewable energy technologies have declined steadily and are projected to drop even more. For example, the average price of a solar panel has dropped almost 60 percent since 2011. The cost of generating electricity from the wind dropped more than 20 percent between 2010 and 2012 and more than 80 percent since 1980. In areas with strong wind resources like Texas, wind power can compete directly with fossil fuels on costs. Cost of the renewable energy will decline even further as markets mature and companies increasingly take advantage of economies of scale.

While renewable facilities require upfront investments to build, once built they operate at very low cost and for most technologies, fuel is free. As a result, the renewable energy prices are relatively stable over time. UCS's analysis of the economic benefits of a 25% renewable electricity standard found that such a policy would lead to 4.1 percent lower natural gas prices and 7.6% lower electricity prices by 2030.

In contrast, fossil fuel prices can vary dramatically and are prone to substantial price swings. For example, there was a rapid increase in U.S. coal prices due to rising global demand before 2008, then a rapid fall after 2008 when global demands declined. Likewise, natural gas prices have fluctuated greatly since 2000.

Extracting Oil and Natural Gas

Extracting oil and natural gas from deposits deep underground is not as simple as just drilling and completing a well. Any number of factors in the underground environment including the porosity of rock and viscosity of the deposit can impede the free flow of product into the well. In the past, it was common to recover as little as 10 percent of the available oil in a reservoir, leaving the rest underground because the technology did not exist to bring the rest to the surface. Today, advanced technology allows production of about 60% of the available resources from a formation.

- Primary recovery: First relies on underground pressure to drive fluids to the surface. When the pressure falls, artificial lift technologies, such as pumps, are used help bring more fluids to surface. In some situations, natural gas is pumped back down the well underneath the oil. The gas expands, pushing the oil to the surface. Gas lift technology is often used in offshore facilities. Primary recovery often taps only 10% of the oil in a deposit.

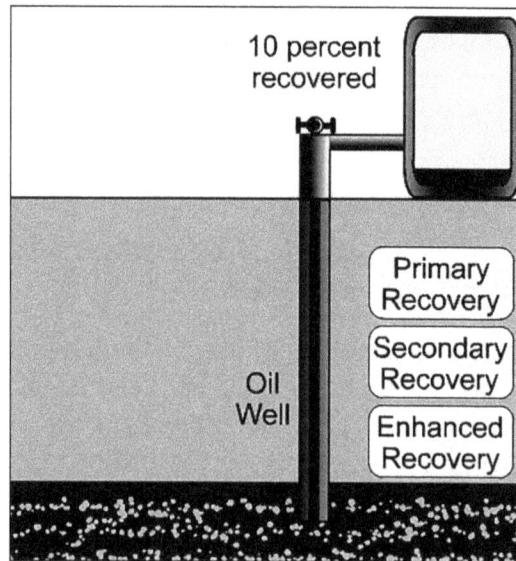

- Secondary recovery is the most widely applied enhanced recovery technique. Water that is produced and separated from the oil in the initial phase of drilling is injected back into the oil-bearing formation to bring more oil to the surface. In addition to boosting oil recovery, it also disposes of the wastewater, putting it back where it came from. This can bring an additional 20% of the oil in place to the surface.

- Enhanced recovery: Techniques are used to mobilize the remaining oil. There are three common approaches: thermal recovery, gas injection or chemical flooding.

- Therma, recovery: Entails injecting steam into the formation. The heat from steam makes the oil flow more easily and the increased pressure forces it to the surface.

- Gas injection: uses either miscible or immiscible gases. Miscible gasses dissolve CO_2, propane, methane or other gasses in the oil to lower its viscosity and increase flow. Immiscible gasses do not mix with the oil, but increase pressure in the "gas cap" in a reservoir to drive additional oil to the well bore.

Chemical flooding involves mixing dense, water-soluble polymers with water and injecting the mixture into the field. The water pushes the oil out of the formation and into the well bore.

Enhanced recovery techniques are employed to bring as much as 60% of the reserve to the surface.

For well over a century, oil and gas have been extracted from underground sources, both on land and increasingly during the past half-century, offshore. The methods of extraction have become very well established and only in the last decade or two has it become necessary to search more vigorously for other resources. The consequences of this extension is that oil is being extracted from different sources - oil shale's and tar sands - while it is now economic to derive natural gas from unconventional sources, with shale gas the most prominent.

As a result, the range of separation process requirements for oil and gas extraction now includes:

- Preparation of drilling muds.

- Separation of solid particles, gases and other liquids from underground crude oil.

- Purification of underground natural gas free from particles, liquids (mainly oil and water) and other gases.

- Preparation of fluids (mainly water) for use in flooding partially exhausted reservoirs to increase oil production.

- Separation of liberated oil from sand and shale deposits and from the liberating water.

- Preparation of the hydraulic fracturing (fracking) liquids for use in the liberation of natural gas from gas shale's.

- Clarification of gases (usually carbon dioxide) prior to their injection into gas reservoirs, to increase gas production rates (or increasingly now, as a means of carbon storage).

- Treatment of produced water, from oil and gas wells (either formation water, formed with the oil or injection water, injected to increase production rates) - to render the produced water fit for reuse or local disposal.

- Production of drinkable water for production sites that is remote from civilization (on shore or off).

- Treatment of all waste waters, including those derived from housekeeping duties, for removal of hydrocarbons and other toxic materials, prior to disposal on land or in the sea.

The above treatment processes are intended either to produce a fluid (gas or liquid) ideally suited to aid in the extraction of fuel material from its reservoir, or to make that material suitable for long distance transport or for local refining or to prepare a waste for safe local disposal.

The production processes are equally dependent on whether they are dealing with oil or gas. Basic oil production involves the drilling of wells to reach the oil-bearing rock formations, followed by the extraction of the oil in one of three successive ways dependent upon the age of the well:

- Primary, which uses reservoir pressure or down-hole pumps to bring the oil to the surface (the deeper the oil-bearing formation, the more likely that some form of artificial lift will be needed).

- Secondary, which uses water flooding or pressurizing of the cap of gas above the oil, to force out more oil.

- Tertiary, usually called Enhanced Oil Recovery (EOR), which involves several different injection processes, using gas, chemical, hot water or steam, or microbes.

The EOR processes all require carefully filtered fluids - carbon dioxide, dilute solutions of alkalis or surfactants, hot water, etc. - to enable them to penetrate the rock formations easily. Even by the use of EOR the recovery of oil from a particular reserve may be no greater than 60 percent of its original content.

This three-stage production process applies to liquid oil trapped below ground, which can be pumped to the surface. A significant proportion of the world's total reserves of hydrocarbon fuels (perhaps as much again as the current total proved liquid oil reserves) exists as the less conventional tar sands, oil shale's and shale gas.

Tar sands take the form of a physical mixture of heavy oil and bitumen with sand. If the deposits are at or close to the surface, as many are, then open-cast mining is used, followed by the hot water processing and flotation to release the oil. For subterranean deposits, recovery is achieved by hot water injection.

In oil shale's, the heavy oil is chemically combined with the rock and requires a chemical separation process to release the oil, such as pyrolysis or hydrogenation. The separation releases vast quantities of waste solids, perhaps yielding 0.2 m^3 of oil per tons of rock processed.

Considerable excitement is currently being generated over the possibility that shale gas might represent a solution to global energy worries. Natural gas is held in shale formations simply because the pores in the rock are too small to allow sufficient mobility. The key to its release is hydraulic fracturing of rock, coupled with the ability to drill horizontally for considerable distances through the shale formations. The hydraulic fluid, which is mainly water, is formulated with a quantity of fine solid particles, which remain in the rock's fissures after fracturing, keeping them open for gas flow. Recognized problems are the possibility of generating earthquakes and the difficulty of preventing methane leakage into deep aquifers and out of the shale area into atmosphere, thus accelerating global warming.

Drilling Muds

In the initial drilling of a well, exploratory or production, the drill bit is surrounded by a thick liquid suspension of clay-like materials-the drilling mud, which is an essential part of the overall oil production process and has important applications for filtration-in their initial formulation and in their recycling, to remove rock fragments. Drilling muds are a complex mixture, so it makes sense to recycle them as much as possible. This creates an exacting task for the mud recycle filter - to remove the maximum of rock fragment content of the returned mud, while changing the basic composition as little as possible.

Oil and Gas Recovery Flows

The most important liquid flow is that of the crude oil being transported up the well bore for immediate treatment. This is carried by flow of produced water, at least as large as the oil flow and generally much larger. There will frequently, also, be a flow of associated natural gas from the oil reservoir.

If the well is in secondary phase production, there will need to be a large water flow necessary for the water flood injection. In tertiary phase, the various enhanced oil recovery processes will also have their operating fluid flows, especially for gas injection, with carbon dioxide flows increasingly important as part of the carbon sequestration schemes required combating global warming.

However the production of oil from tar sands is, on the global scale nowhere near that of pumped crude oil, its fluid flow rates will become large, as the reserves are developed. These will include, besides the produced oil, residual water from the processing of open-cast sands or the recovery of oil from underground. Similar flows will occur in shale oil production.

Crude Oil Treatment

The recovery of crude oil from underground requires separation treatment in two main places: at the well bottom and at the well head. In the very restricted space at the bottom of the producing well, solid/liquid filtration is necessary to prevent the passage up the well pipe of as much suspended solids as possible. This is done by well screen, a zone of perforated material that is either built in to the end of the well pipe or fitted as a sleeve over a very coarsely perforated part of the pipe. The well screen is a specialized form of filter. It can be made from wire mesh, perforated plate or porous metal fiber material.

Once the oil reaches the surface, there is more working space for any required filtration and the major separation requirement is to recover the crude oil from its mixture with produced water. This is very often undertaken in liquid/liquid separators working by sedimentation, almost certainly in the lamellar separators for off-shore installations where not so much space is available. Production economies dictate that this separation should be as efficient as possible, since the separated water may be going to waste, carrying any unseparated oil with it. This means that any emulsion of oil and water will have to be broken before final oil/water separation. For safety's sake, a micro-filter may be used at the well head, although the flow rates will be high and filters will have to be automatically or easily manually cleaned.

Gas Treatment

Natural gas produced in association with crude oil will not normally present a filtration problem - from solids at least, however it may need separation from oil or water droplets. However, there is nowadays an increasing need for the injection of gases into underground strata, to improve oil production rates. Increasingly this is done directly into the rock formation, either as an enhanced oil recovery process or as a sequestration method for carbon dioxide disposal.

Direct injection of gases will require that they be free from suspended solids, possibly down to the same size level as is the case for water injection, namely around 2μm. This will be done in the same sort of filters as are used for engine intakes, using V-block minipleat filter panels, for example.

Produced Water Treatment

The associated (produced) water is made up from the layer of water that is normally found under the oil layer in the reservoir, together with any injection water used to boost oil or gas production and so increases, with time, as a proportion of the total fluid extracted. In chemical terms this flow of water is a very complex mixture of organic and inorganic materials, both dissolved and suspended in the water flow. As it is normally intended to dispose of this water in vicinity of the well (particularly for offshore wells), this complexity presents a serious task to the well operator.

The quality of produced water as it leaves the well site depends upon where that site is: onshore or offshore, shallow sea or deep sea, close to a town or far from human habitation. In its untreated state, produced water is more saline than seawater and so cannot be put back in the sea without significant disturbance of the marine environment. Neither can it be discharged onto land before its salt content has been reduced.

It is possible that the produced water can be used, after suitable treatment, for injection, back into the formation to promote oil flow. This water will be injected into an underground reservoir and must be able to flow through the small passageways in the rock. This means that it will have to be filtered free of fine solids, possibly down to 2μm at the point of injection (however there will not be the need to separate oil from it so thoroughly). Where there is the space so to do, this filtration can be achieved by the deep bed ('sand') filters, almost certainly using multimedia beds for most efficient operation. For large operations, a two-stage filtration may make sense, with a 10μm limit for water flow through topside equipment and further filtration down to 2μm or less at the point of injection.

Other Water Flows

Oil and gas wells tend to be in isolated places, offshore or otherwise well away from mains water supply. Human needs for fresh drinkable water are normally met by desalination, since its high energy cost can be sustained at the well site.

The increasing need for ballast water processing, to prevent long-distance transport of marine flora and fauna, has led to the development of special centrifugal equipment for this purpose.

A very different water treatment need exists in the production of oil from tar sands. Here the water is used, hot, to process the sands, to yield effluent water high in sand content. There is no equivalent to the water flooding of crude oil reservoirs, so there is a use for this water only when it can be recycled to the main production process. Filtration will be needed in this system, however mainly to protect the operating equipment.

Technology Trends

The major separation process in the oil and gas extraction business is undoubtedly that of oil from water, with the purification of gas second in importance mainly because it is easier. These two situations mean that the bulk of separations technology in the industry is given over to sedimentation equipment, with knock-out drums and API and lamellar separators most important. On the whole, this is a very mature part of the filtration business, using very well established forms of equipment, although demands on space are increasing the use of centrifugal oil/water separators, especially on offshore rigs, where energy costs are relatively low.

However, some changes are discernable, with the most important being driven by the needs for greater separation efficiency and finer cut points in most separations. The most apparent change is in the use of the membrane systems, MF and UF in particular, for the removal of the last traces of oil in the effluent from waste treatment plants. The membranes are also finding application, especially for MF systems, in the preparation and recycling of injection fluids and of fluids for fracking.

Coarser filtration is finding applications in the extraction of hydrocarbons from tar sands.

2.6   Land Resources

The occupation is nearly 20% of the earth surface, measuring almost up to about 13000 million hectares of area. The houses, roads and factories occupy nearly one third of the land. The forests occupy another one third of the land.

The rest of land is used for and for meadows and pastures. The soil forms the surface layer of the land which covers more than the 80 percent of land. The soil is defined as a natural body which keeps on changing and even allows the plants to grow.

It consists of organic and inorganic materials. This definition is given by Buckman and Brady. The branch of science which can deal with the formation and distribution of soil in the different parts of the world is referred as a pedology. The professional which deals with the soil is known as the pedologist.

The inorganic component in the soil is 45 percent and the organic component in the soil is 5%. The water component in the soil is 25% and air component in the soil is 25%. The soil particles have fine spaces which are known as pore spaces. These are also known as the interstices.

They contain air and water along with the dissolved substances. The water and air content in the soil is inversely related to each other. The more is the water content lesser is the space for air to exist. The soil has both animals and plants. The micro flora consists of heterotrophic and autotrophic bacteria.

It also contains the fungi and algae. The heterotrophic bacterium consists of nitrogen and non-nitrogen fixing bacteria. Nitrogen fixing bacteria can be symbiotic, non-symbiotic, aerobic and anaerobic. Non nitrogen fixing bacteria can either be aerobic or anaerobic. The fungus includes yeast and mushrooms.

Algae can be red or brown or green. The fauna can be micro or macro. The micro fauna includes protozoa and nematodes. The macro fauna includes the mites, termites, earthworm, snails and mice. The soil has different types of soil particles. The mineral composition of the rock determines them along with the size of particles.

It includes gravel particles, sand, silt and clay particles. The gravel particles consist of mainly small stones and have a few sand particles and are used to make roads. The sand particles consist of pores and are aerated. They can hold little bit of water and are made up of large quartz. The silt particles are moved by the help of water. They are left at the bank of river. They are inert and are made up of large quartz.

The clay particles contain nutritive salts and have ability to retain the water. They are not inert and react chemically. Some of their pure forms are not suitable for growth of plants as they form a non-penetrable mass. The other components of soil mix with the clay particle and form a granular soil. This type of soil is ideal for the cultivation. It contains pores as well as has the ability to hold water. It also contains nutritive salts.

The loamy soil is made up of clay, silt and sand. The proportion of clay is least and is half as compared to the silt and sand. The silt and the sand are twice and equal in the proportion. It is also a good soil for the growth of plants as it has pores as well as has the ability to hold water. It also contains some nutritive salts.

There are many factors which control the nature of soil. They are porosity, water holding capacity and texture. They come under the physical nature of soil. The chemical nature of the soil is governed by the salt content, inorganic and organic content includes certain metals. The climate, topography and organisms also play a vital role in deciding the nature of soil. The half decayed and half synthesized part of organic material in the soil forms the humus. It contains nutrients and help in growth. It has the ability to absorb the heat and warm the soil. It makes the soil granular by its porosity and water holding capacity.

2.6.1 Land Degradation

Land degradation is the process of deterioration of soil or loss of fertility of the soil. It is the reduction in capacity of the land to provide ecosystem goods and services and assure its functions over a period of time. Land degradation affects large areas and many people in the dry land regions. Increased population pressures and excessive human expansion into dry lands during long wet periods leave an increasing number of people stranded there during dry periods. The transfer of critical production elements to other uses through the introduction of irrigated and non-irrigated cash crops and the use of water for the industrial and urban purposes at the expense of rural agricultural producer break links in the traditional production chains in dry lands. The removal of protective cover to reduce the competition for ploughing, water and nutrients, heavy grazing and deforestation leave the soil highly vulnerable to wind erosion particularly during severe droughts. Heavy grazing around water points or during long droughts prevents the regrowth of vegetation or favors only unpalatable shrubs.

Harmful effects of land (soil) degradation:

- The soil texture and soil structure are deteriorated.

- Increase in water logging, salinity, alkalinity and acidity problems.

- Loss of soil fertility, due to loss of invaluable nutrients.

- Loss of economic social and biodiversity.

Causes of Land Degradation

Population:

As population increases, more land is needed, for producing food, fiber and fuel wood. Hence there is more and more pressure for the limited land resources, which are getting degraded due to over exploitation.

Urbanization:

The increased urbanization due to population growth reduces extent of agricultural land. To compensate the loss of agricultural land, new lands comprising natural ecosystems such as forests are cleared. Thus the urbanization leads to deforestation, which in turn affects millions of plant and animal species.

Fertilizers and Pesticides:

Increased applications of fertilizers and pesticides are needed to increase farm output in the new lands, which again leads to pollution of land and water and soil degradation.

Damage of Top Soil:

Increase in soil food production generally leads to damage of top soil through nutrient depletion.

Water Logging:

Soil erosion, desalination and contamination of the soil with Industrial wastes all cause land degradation.

2.6.2 Wasteland Reclamation

Wasteland

The land which is not in use is called waste land. The waste land is unproductive, unfit for cultivation, graying and other economic uses. About 20% of the geographical area of India is waste land.

Types of Wastelands

Wastes lands can be divided into two types. They are as follows:

- Uncultivated waste lands.
- Cultivable waste lands.

Uncultivated Waste Lands:

These lands cannot be brought under cultivation.

Example: Barren rocky areas, hilly slopes, stony or leached or gully land or sandy deserts, etc.

Cultivable Waste Lands:

These are cultivable but not cultivated for more than five year. Cultivable waste lands are important for agricultural purposes.

Examples: Degraded forest lands, gullied water, logged and marsh lands, saline lands, etc.

Causes of waste land formation:

- Due to soil erosion, deforestation, over graying, water logging, salinity.

- The increasing demand for firewood and exercise use of pesticides.

- Over exploitation of natural resources.

- By the sewage and industrial wastes.

- Mining activities destroy the forest and cultivable land.

Objectives or Need of Waste land reclamation:

- To improve the physical structure and quality of the soil.

- To prevent soil erosion, flooding and landslides.

- To avoid over exploitation of natural resources.

- To improve the availability of good quality of water for agricultural purposes and industrial operations.

- To conserve the biological resources and natural ecosystem.

- To provide a source of income to the poor.

Methods of Waste Land Reclamation

Drainage:

Excess water is removed by artificial drainage. This process is used for water logged coil reclamation.

Leaching:

Leaching is the process of removal of salt from the salt affected soil by applying access amount of water. Leaching is done by dividing the field in small plots. In continuous leaching 0.5cm to 1.0cm water is required to remove 90% of soluble salts.

Irrigation practice:

High frequency irrigation with controlled amount of walls helps to maintain better water availability in the Land.

Application of Gypsum:

Soil can be reduced with gypsum. Calcium of gypsum replaced sodium from the unchangeable sites. This process converts clay back into calculus clay.

Social Forestry Programme:

These programmers induce string plantation on road, land sides, degraded forest land, etc.

2.6.3 Man Induced Landslides

The causes of man induced landslides are as follows:

- Excavation.
- Loading.
- Draw-down.
- Land use (e.g., construction of roads, houses etc.).
- Water management.
- Mining.
- Quarrying.
- Vibration.
- Water leakage.
- Deforestation.
- Land use pattern.
- Pollution.

Human Factors in Man Induced Landslides

These are basically the human activities like construction of roads, buildings, dams, etc. These have strong bearing on the Man Induced Landslides. First road in Himalayan region was introduced way back in British colonial period (Singh and Ghai, 1996). It was, however, only after the Indo-China war of 1962 that the road construction was intensified in the Himalayas.

Thereafter, Indian engineers blasted gigantic networks of road and communication facilities deep into hills of the Himalaya. In NDBR, the history of road construction began in sixties. In 1964, for first time, roads were constructed in the reserve. Local people reported that initially about 80 km of roads were constructed in the region. About 50 km of road were further added to the existing network of roads in 2000.

Presently, road length is about 135 km Now the road in Mana valley has been declared as national highway (Haridwar-Badrinath National Highway) due to which slopes are now in process of being over-modified. Thus, sheer stress factors are exceeding sheer strength factors and landslides are increasing. After the introduction of roads, landslides have become very frequent in the reserve. The correlation between construction of roads and landslides is positive. It has the three stage relations.

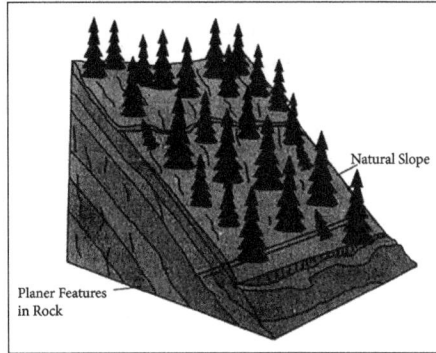

Stage I: Natural slopes - Premodified stage.

First stage, which is the premodified stage of natural slope, the steady-state condition is found. Then, the natural slopes are modified or undercut for the construction of roads and dams, etc. Thus, steady-state condition is disabled.

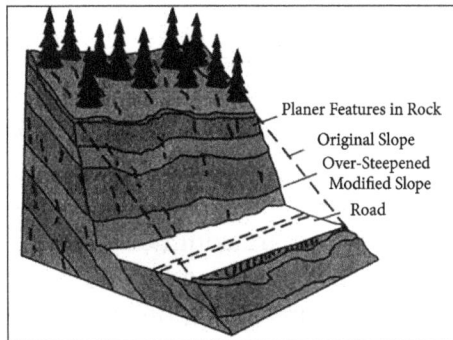

Stage II: Disabled steady-state condition.

Then landslides take place so that steady-state condition is reestablished. Therefore, landslides can be said nature's rule of equilibrium.

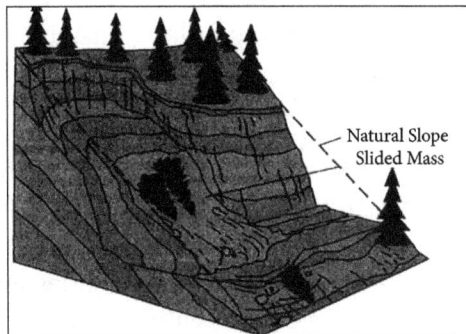

Stage III: Nature Re-establishing steady-state condition.

Built-up land has also increased in the recent past due to which natural slopes have been over-modified. It has also led to deforestation in the reserve. Thus, steady-state condition has disabled in many areas, which have eventually resulted in increased frequency of landslides.

Strong positive correlation between occurrence of landslides and construction of roads were observed during the field investigations. Villagers reported that earlier landslides were not very common but after the introduction of roads, this has become a common phenomenon. About 80 percent of the total landslides occurring now in the reserve are a result of road construction.

According to field survey, most of villagers consider that road construction excessive rainfall and seismicity are main causes of landslides. About 83.5%, 88% and 88.5% of the respondents reported road construction, excessive rainfall and seismicity respectively as the most common causes of landslide occurrence in the reserve. Whereas declining forest cover, grazing and agricultural land expansions are other causes of landslides.

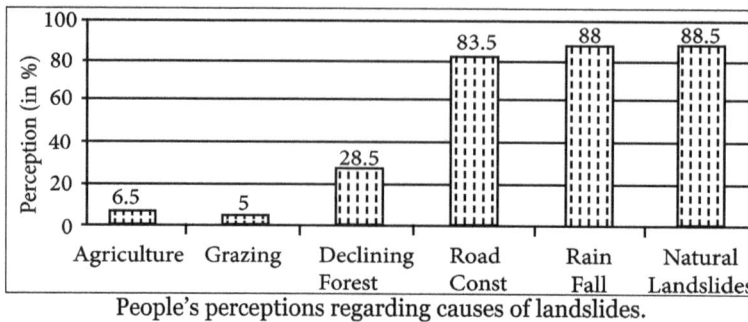

People's perceptions regarding causes of landslides.

Agriculture, grazing and deforestation are commonly regarded as the main causes of landslides in other Himalayan regions but these were reported as minor causes of landslides in the reserve, which is the result of ban on utilization of local resources in reserve. Road construction has been the prime cause of landslides in the region.

Most of the region is located in the peri-glacial and glacial environment. So, ideally, there should be much snowfall rather than the rainfall. But local people reported that snowfall has decreased and rainfall has increased in last few decades. This can be attributed to global warming due to which rainfall regions are shifting upwards and taking place of snowfall in the reserve. Thus, excessive rainfall in the first half of monsoon saturates the bedrock and in latter half removes it in the form of landslides.

Road construction and rainfall have increased in the reserve thus landslides have also become very frequent. Survey results show that the landslides have been increasing rapidly in region since last three decades, as about 51 percent of the respondents reported the same, while 33 percent reported that landslides are increasing slowly and only 16 percent stated that trend of landslide is almost static.

Historical trend of major landslides in reserve has been constructed on the basis of literature and field survey. A total of 10 major landslides have been noticed in literature. There must be some more massive landslides in the region.

However, these are not properly documented due to the fact that most of the reserve is highly inaccessible to the outer world. Thus, most of the landslides of the reserve remain unregistered.

Table: History of Landslides in NDBR.

Year	Causes	Description
1939	-	Man village was partially abandoned due to extensive rock falls.
1978	Avalanches	Bamni village near Badrinath Puri was completely washed away.
1968	Related to Flash Floods	Rishi Ganga in Garhwal was blocked up to a height of 40 m due to a slide at Reni village. The dam breached in 1970 caused extensive damage.
July 1970	Exceptionally Heavy Rainfall	Alaknanda river caused considerable loss of life among pilgrims. Many bridges. Houses and an entire village were washed away.
September 1970	Exceptionally Heavy Rainfall	Landslide and house collapses killed 223 people.
July 1970	landslide due to Dams	Floods in Rishi Ganga created 40m high blockade near village of Reni in U.P. Lake silted up by May. 1970 and eventually blockade breached in July, 1970.
September 1999	Earthquake-induced landslide	North of Joshimath

During field survey, about 40 landslides were noticed in reserve. Of the total landslides, about 31 were human-induced and 9 were natural. All the human-induced landslides have recent origin. Thus, it can be said that human activities have aggravated the problem of landslides only in recent past.

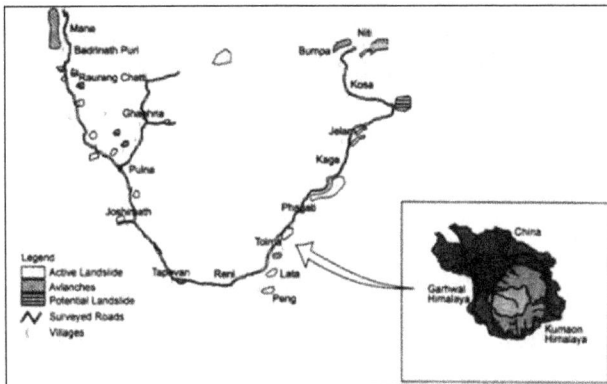

Landslides in nanda devi bio-sphere reserve.

Presently, intensity of landslide occurrence is high in NDBR as about 87 percent of the respondents indicated the same, whereas only 13% indicated moderate, low and very low intensity. People who reported moderate, low and very low intensity of landslides inhabit upper reaches of river valleys and practice seasonal migration.

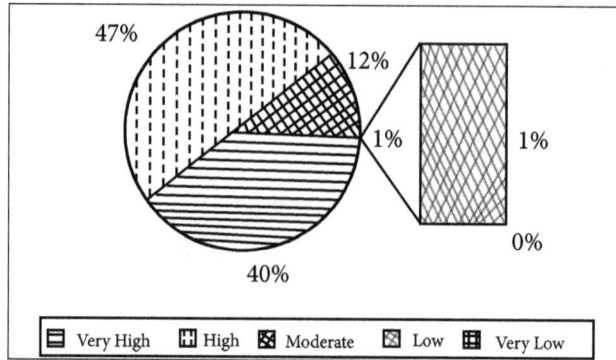

People's perceptions regarding intensity of landslides.

During monsoon, upper reaches receive less rainfall, thus landslides are not as common as in lower reaches of the river valley. During winters, when avalanches take place, inhabitants of upper reaches migrate to low altitude and do not witness cases of frequent landslides.

2.6.4 Soil Erosion and Desertification

Soil is naturally created when the small pieces of weathered rocks and minerals mix with the organic materials from decaying plants and animals. Soil creation is a slow process, taking many years. However, the soil that is created is constantly subjected to natural and manmade forces that disrupt it.

Soil erosion is a naturally occurring process that affects all the landforms. In agriculture, soil erosion refers to the wearing away of field's topsoil by the natural physical forces of water and wind or through the forces associated with farming activities such as tillage.

Causes of Soil Erosion

Erosion occurs when farming practices are not compatible with the fact that soil can be washed away or blown away. These practices are as follows:

- Inappropriate farming techniques such as deep ploughing of land 2 or 3 times per year to produce the annual crops.

- Overstocking and overgrazing.

- Planting the crops down the contour instead of along it.

- Lack of crop rotation.

Erosion, whether it is by water, wind or tillage, involves three distinct actions such as soil detachment, movement and deposition. Topsoil, which is high in organic matter, fertility and soil life is relocated elsewhere "on-site" where it builds up over time or is carried "off-site" where it fills in the drainage channels. Soil erosion reduces the cropland productivity and contributes to the pollution of adjacent wetlands, watercourses and lakes.

Soil erosion can be a slow process that continues relatively unnoticed or can occur at an alarming rate, causing serious loss of topsoil. Soil compaction, loss of soil structure, low organic matter, salinization, poor internal drainage and soil acidity problems are other serious soil degradation conditions that can accelerate the soil erosion process.

Soil erosion is defined as the wearing away of topsoil. Topsoil is the top layer of soil and is the most fertile because it contains the most organic, nutrient-rich materials. Hence, this is the layer that farmers wants to protect for growing their crops and ranchers want to protect for growing grasses for their cattle to graze on.

One of the main causes of soil erosion is water erosion, which is the loss of topsoil due to water. Raindrops fall directly on the topsoil. The impact of the raindrops loosens the material bonding it together, allowing small fragments to detach. If the rainfall continues, water gathers on the ground, causing water flow on the land surface which is termed as surface water runoff. This runoff carries the detached soil materials away and deposits them elsewhere.

There are some conditions that can accentuate the surface water runoff and therefore, soil erosion. For example, if the land is sloped, there is a greater potential for soil erosion due to gravity which pulls the water and soil materials down the slope.

Effects of Soil Erosion

The loss of natural nutrients and possible fertilizers directly affect the crop growth, emergence and yield. Seeds can be disturbed or removed and pesticides can be carried off. The soil structure, quality, texture and stability are also affected, which in turn affects the holding capacity of the soil.

Eroded soil can inhibit the growth of seeds, bury seedlings, contribute to road damage and even contaminate the water sources and recreational areas.

Desertification

Desertification is a progressive destruction or degradation of arid or semi-arid lands to desert. It is also a form of land degradation. Desertification loads to the conversion of range lands or irrigated croplands to desert like conditions in which agricultural productivity falls. Desertification is characterized by denegation, depletion of ground water, desalination and soil erosion.

Causes of Desertification:

- Deforestation: The process of deserting and degrading a forest land initiates a desert. If there is no vegetation to hold back the rain water, soil cannot soak and ground water levels do not increases. This also increases, soil erosion, loss of fertility.

- Overgrazing: The increase in cattle population heavily grazes the grass land or forests and as a result denude the land area. The denuded land becomes dry, loose and more prone to soil erosion and leads to desert.

- Pollution: Excessive use of fertilizers and pesticides and disposal of toxic water into the land also leads to desertification.

Harmful Effects of Desertification:

- Around 80% of the productive land in the arid and semi-arid regions are converted into desert.

- Around 600 million people are threatened by desertification.

2.6.5 Role of an Individual in Conservation of Natural Resources

Since resources are being exhaustible, it is the duty of every individual on this earth has to conserve the natural resources in such a way that they must be available for future generation also. Individual must understand the essential if natural resources. Due to advancement in technology and population growth, the present world is feeling lot of problems on degradation of natural resources.

Measures recommended for conservation of natural resources:

Conservation of Energy:

- Switch off lights, fans and other appliances when not in use.

- Use solar heater for cooking our food on sunny days, which will cut down our LPG expenses.

- Dry the clothes in sunlight instead of driers.

- Grow trees near the houses and get a cool breeze and shade. These will cutoff our electricity charges on A/C and waters.

- Use always pressure cooker.

- Ride bicycle or just walk instead of using car and scooter.

Conservation of Water:

- Use minimum water for all domestic purposes.

- Check for water leaks in pipes and toilet and repair them promptly.

- Use drip irrigation to improve the irrigation efficiency and reduce evaporation.

- Reuse the soapy water, after washing clothes, for washing off the country yards, drive ways, etc.

- Build rainwater harvesting system in our house.

- The wasted water, coming out from kitchen, bath tub, can be used for watering the plants.

Conservation of Soil:

- While constructing the house don't uproot the trees as for as possible.

- Grow different types of plants, herbs, trees and grass in our garden and open areas, which bind the soil and prevent its erosion.

- Soil erosion can be prevented by the use of sprinkling irrigation.

- Don't irrigate the plants using a strong flow of water, as it will wash off the top soil.

- Uses mixed cropping, so that some specific soil nutrients will not get depleted.

- Use green manure in the garden, which will protect the soil.

Conservation of Food Resources:

- Eat only minimum amount of food. Avoid over eating.

- Don't wastes the food instead gives it to someone before getting spoiled.

- Look only required amount of the food.

- Don't cook food unnecessarily.

- Don't store large amounts of food grains and protect them from damaging insects.

Conservation of Forest:

- Use non-timer products.

- Plant more trees and protect them.

- Grassing, fishing must be controlled.

- Minimize the use of papers and fuel wood.

- Avoid of executing developmental work like dam, road and construction in forest areas.

2.6.6 Equitable use of Resources for Sustainable Lifestyles

Scarcity of the resources is the burning problem of modern technology. The twenty first century will see growing human needs for resources as many parts of the world are using natural resources at a rate faster than the natural processes can recharge it.

Natural resources are limited. For example, an existing water sources are being subjected to heavy pollution. Global climatic changes are changing the quality of fresh water sources as a consequence of the unknown effects on the hydrological cycle.

Sustainable development is currently being a field of development, planning and other associated aspects. In the light of the self-defeating current mode of development and recurrent natural calamities, people are urged to ponder over the faults, lacunae, short-comings, discrepancies and limitations of the ongoing process and production system.

It is an essential to sustain the natural resources. We should conserve the natural resources so that it may yield sustainable benefit to the present generation while main-taining its potential to meet the needs of the future generation. There are three specific objectives to conserve living resources:

- To preserve the biodiversity.

- To ensure that any utilization of the ecosystem is sustainable.

- To maintain the essential ecological processes.

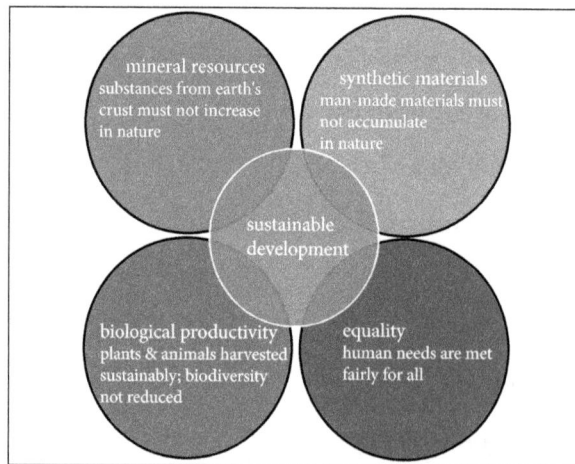

Resource management should be less energy-intensive, less cost-intensive and more viable in terms of economy, ecology and culture, suitable to local ecology and needs of the people. The Srilankan team, for example studied traditional paddy irrigation sys-tems as a model for water management.

Its report note that from the 5th century B.C. through the 12th century A.D., Sri Lan-ka developed a technologically advanced civilization based on an intricate system of rainwater conservation and irrigation. Water users were collectively and individually responsible for maintenance of the irrigation systems and customary laws, known as Sirit, were established governing the water use and the related aspects of life.

Similarly, the italics system is a system of farmer-managed canal irrigation, which has been in operation for more than 300 years in Dhule and Nashik district of northwest-ern Maharashtra.

Chapter 3

Biodiversity and its Conservation

3.1   Introduction to Biodiversity

Biodiversity is defined as the variety and variability among all groups of living organism and the ecosystem in which they occurs.

Classification of Biodiversity

Biodiversity is generally classified into three types:

- Genetic diversity.

- Species diversity.

- Community or Ecosystem diversity.

Genetic Diversity

A species with different genetic characteristics are known as subspecies or genera.

Genetic Diversity is Diversity within Species:

Within individual species, there are number of varieties which are slightly different from one another. These differences are due to the differences in the combination of gases. Genes are basic units of hereditary information transmitted from one generation to other.

Examples: i. Rice Varieties

All rice varieties belong to species "Organization". But there are, thousands of rice varieties, which show variation at genetic level differ in their size, shape, color and nutrient content.

ii. Teak Wood Varieties

There are number of teak wood varieties found available.

Example: Indian teak, Burma teak, Malaysian teak, etc.

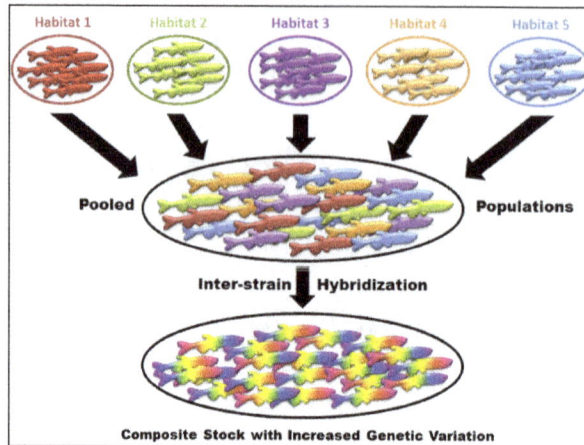

Genetic diversity.

Species Diversity

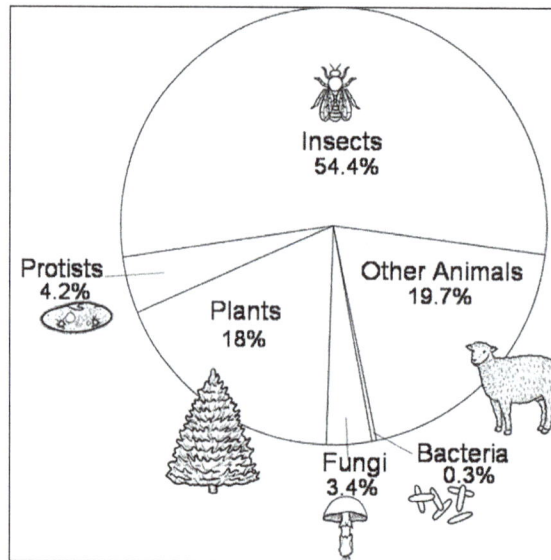

Species diversity Graph.

A discrete group of organisms of the same kind is known as species. Species diversity is the diversity between differed species. The sum of varieties of the entire living organism at the species level is known as species diversity.

Biotic component is composed of a large number of species of plants, animals and microorganisms which interact with each other and with the abiotic component of the environment.

Example:

- Total number of living species in the earth is more than 20 million. But, of which only about 1.5 million living organisms are found and given scientific names.

- Plant species: Apple, mango, grapes, wheat, rice, etc.

- Animal species: Lion, tiger, elephant, dear, etc.

Community Ecosystem Diversity

It is a set of biotic components interacting with one another and with abiotic components.

The diversity at the ecological or habitat level is known as ecosystem diversity. A large region with various ecosystems can be considered as ecosystem diversity.

Example: River ecosystem

The river which include the fish, aquatic insects, muscles and variety of plants that have adapted. Thus, the ecosystem diversity is a aggregate of different environment types in a region. It explains the interaction between living organisms and physical environment in ecosystems.

3.2 Value of Biodiversity: Consumptive use, Productive use and Social Values

Value of Biodiversity

Biosphere is a life supporting system to the human beings. It is the combination of different organisms. Each organism in the biosphere has its own significance. Biodiversity a vital for healthy biosphere. Biodiversity is must for the stability and proper functioning of the biosphere. We get benefits from other organisms in number of ways. Sometimes we realize the real value of the organism only after it is lost in this earth.

The values of biodiversity have been classified into 6 types.

1. Consumptive use value

These are direct use values where the biodiversity products are harvested and consumed directly.

Example: Food, Drug, Fuel, etc.

i. Food

A large number of wild plants are consumed by human beings as food. Nearly 80 to 90% of our food crops have been domesticated only from the tropical wild plants. A large number of wild animals are also consumed as food.

Example:

- In Himalayan region.

- Insects: Spider and wild herbivores are consumed by many tribal and non-tribal Communities in India.

ii. Drugs

Around 70% of modern medicines are derived from plant and plant extracts. 20,000 plant species are believed to be used medicinally, particularly in the traditional system of Ayurveda and Sidha.

Example:

- Germany alone use more than 2,500 Species of plants for medicinal purposes in Homeopathy and other systems of medicines.

- India uses 3000 species of plants in Ayurveda, Homeopathy and System of medicines.

iii. Fuel

Fire woods are directly consumed by villagers or tribals. The fossil fuels like coal, petroleum and natural gas are also the products of fossilized biodiversity.

2. Productive Use Values

Biodiversity products have obtained commercial values. These products are marketed and sold. These products may be derived from the animals and plants.

Animal Product	Animal
Silk	Silk-worm
Wool	Sheep
Leather	All animals
Food	Fish and animals

Table: Plant products for various industries.

Plant Product	Industry
Wood	Paper and pulp industry, plywood industry
Cotton	Textile industry
Fruits, vegetables	Food industry
Leather	Leather industry

- Rice accounts for 22% of the cropped area and cereals accounts for 39% of the cropped area.

- Oil seed production also helped in saving large amount of foreign exchange spend on importing edible oils.

3. Social Values

Social value of biodiversity refers to the manner in which the bio resources are used to the society. These values are associated with the social life, religion and spiritual aspects of the people.

Example:

i. Holy Plants

Many plants are considered as the holy plants in our country.

Example: lotus.

The leaves, fruits of these plants are used in worship.

ii. Holy Animals

Many animals are also considered as holy animals in our country.

Example: Cow, Snake, Bull, Peacock.

4. Ethical Value

It involves ethical issues like all life must be preserved.

In India and in other countries biodiversity is considered to have great value on religions and cultural basis.

Our rich heritage teaches us to worship plants, animals, rivers and mountains. The ethical value, means that a species may or may not be used, but its existence in the nature gives us pleasure.

5. Aesthetic Value

The beautiful nature of plants and animals insist us to protect the biodiversity. The most important aesthetic value of biodiversity is ecotourism.

Example:

- Eco-tourism: People from far place spend a lot of time and money to visit the beautiful areas, where they can enjoy the aesthetic value of biodiversity. This type of tourism is known as ecotourism.

- The pleasant music of wild birds.

- Color of butterfly, color of flowers and color of peacocks are very important for their aesthetic value.

6. Option Values

The option values are the potential of biodiversity that are presently unknown and need to be known. The optimal values of biodiversity suggest that any species may be proved to be a valuable species after some day.

Example:

- The growing biotechnology field is searching a species for causing the disease of cancer and AID's.

- Medicinal plants and herbs play a very important role in our Indian economic growth.

3.3   Biodiversity at National and Local Levels

There are at present 1.8 million species known and documented by scientists in the world. Although, scientists have estimated that the number of species of plants and animals on earth could vary from 1.5 to 20 billion. Thus the majority of species are yet to be discovered.

Biodiversity is the measure of the variety of earth's animal, plant and microbial species of genetic differences within species and of the ecosystems that support the species. Out of an estimated 30 million species on earth, only one-sixth has been identified and authenticated in the past 200 years.

An estimated biodiversity covers 400,000 higher plants. Most of the world's bio-rich nations are in the South, which are developing nations. In contrast, the majority of the countries capable of exploiting biodiversity are Northern nations, in the economically developed world.

These nations however have low levels of biodiversity. Thus the developed world has come to support the concept that biodiversity must be considered to be a 'global resource'. Although, if biodiversity should form a 'common property resource' to be shared by all nations, there is no reason to exclude oil or uranium or even intellectual and technological expertise as global assets.

India's sovereignty over its biological diversity cannot be compromised without a revolutionary change in world thinking about sharing of all types of natural resources. The biodiversity of 89 countries with diversities higher than India are located in South America such as Brazil and South East Asian countries such as Malaysia and Indonesia.

The species found in these countries, however, are different from our own. This makes it imperative to preserve our own biodiversity as a major economic resource. While few of the other 'mega-diversity nations' have developed the technology to exploit their species for biotechnology and genetic engineering, India is capable of doing so.

Throughout the world, the value of biologically rich natural areas is now being increasingly appreciated as being of unimaginable value. International agreements such as the World Heritage Convention attempt to protect and support such areas. India is a signatory to the convention and has included several protected Areas as World Heritage sites.

These include Manas on the border between Bhutan and India, Kaziranga in Assam, Bharatpur in U.P., Nandadevi in Himalayas and the Sunderbanns in the Ganges delta in West Bengal. India has also signed the Convention in the Trade of Endangered Species (CITES) which is intended to reduce the utilization of endangered plants and animals by controlling trade in their products and in the pet trade.

Biologically, tropical rain forests are the centres of the world much of the earth's contemporary flora and fauna originated in the humid tropics. For millions of years, tropical rain forests have been factories of evolutionary diversity from which plants and animals, capable of adapting to more difficult environments, have gone forth to populate the subtropical and temperate regions. It is essential to maintain areas of tropical rain forest large enough for this evolution to continue. The tropical forests are regarded as the richest in biodiversity. The species diversity in tropics is high.

The reasons are as follows:

- Warm temperate and high humidity provide favorable conditions for many species.

- Tropical communities are more productive because these areas receive more solar energy.

- Over geographical times the tropics have had a more stable climate. In tropics, therefore, local species continued to live there itself.

- Among plant rates of out crossing appear to be higher in tropics.

The biodiversity exists on earth in eight broad realms with 193 bio-geographical provinces. Each bio-geographical province is composed of the ecosystems, which are constituted by communities of living species existing in an ecological region.

The developing countries, located in subtropical/tropical belt are far richer in biodiversity than the industrial countries in the temperate region. The Valvilovian Centres of diversity of crops and domesticated animals are also located in the developing countries.

It is important to preserve the numerous varieties of plants and animals that belong to one species. Each variety within a species contains unique genes and the diversity of genes within a species increases its capacity to adapt to pollution disease and other changes in the environment.

3.3.1 India as a Mega-diversity Nation

Among the biologically rich nations, India stands among the top 10 or 15 countries for its great variety of plants and animals, many of which are not found elsewhere. India has 350 different mammals, 1,200 species of birds and 453 species of reptiles and 45,000 plant species, of which most are angiosperms.

These include generally high species diversity of ferns and orchids. India has 50,000 known species of insects, which includes 13,000 butterflies and moths. It is estimated that the number of unknown species could be several times higher. It is estimated that 18% of Indian plants are endemic to country and found nowhere else in the world.

As one of the oldest and largest agriculture societies, India has also a striking variety of at least 166 species of crop plants and around 320 species of wild relatives of cultivated crops. There is a vital, but often neglected factor when we focus on biodiversity. It may be a matter of she surprise for many to understand that the who officially constitute 7.5 percent of India's population have preserved about 90 percent of the country's diversity. To a large extent, survival of our biodiversity depends on how best they are looked after.

3.3.2 Hot-spots of Biodiversity

The hot spots are the geographic areas which possess the high endemic species.

At the global level, these are the areas of high conservation priority. If these species lost, they can neither be replaced nor regenerated.

Criteria for recognizing hot spots:

- The hot spots should have a significant percentage of specialized species.
- The richness of the endemic species is the primary criterion for recognizing hot spots.
- The site is under threat.
- It should contain important gene pools of plants of potentially useful plants.

Reasons for Rich Biodiversity in the Tropics

The following are the reasons for the rich biodiversity in the tropics:

- Warm temperatures and high humidity in the tropical areas provide favorable conditions.
- The tropics have a more stable climate.

- Among plants, rate of outcrossing appear to be higher in tropics.

- No single species can dominate and thus there is an opportunity for many species to coexist.

Hot spots of Biodiversity in India:

- Western Ghats - Sri Lanka Region.

- Eastern Himalayas - Indo-Burma Region.

Eastern Himalayas: Geographically this area comprises of Nepal, Bhutan and neighboring states of Northern India. There are 35,000 plant species found in the Himalayas, of which 30% are endemic.

The Eastern Himalayas are also rich in wild plants of economic value.

Examples: Rice, Banana, Juice and Sugarcane.

The taxol yielding plant is also sparsely distributed in the region:

- 63% mammals are from Eastern Himalayas.

- 60% of the Indian Birds are from North West.

- Huge wealth of fungi, insects, mammals, birds have been found in this region.

Western Ghats: The area comprises Maharashtra and Kerala. Nearly 1500 endemic, dicotyledons plant species are found from Western ghats. 62% amphibians and 50% lizards are endemic in Western Ghats. It is reported that only 6.8% of the original forests are existing today while the rest has been deforested or degraded.

Some common plants: Japonica, Rhododendron and Hypericum.

Some common animals: Blue bird, Lizard, hawk.

3.4   Threats to Biodiversity: Habitat Loss and Man-Wildlife Conflicts

Extinction is the Complete Elimination of Wild Species

It is a Natural but slow process but due to unplanned activities of man; the rate of decline of wild life has been particularly rapid in the last one hundred years. There are a number of causes which are known to cause the extinction of wildlife.

Wetlands have been drained to increase agricultural land. These changes have economic implications in the longer term. The current destruction of the remaining large areas of wilderness habitats, especially the super diverse tropical forests and coral reefs, is the most important threat worldwide to biodiversity. Scientists have estimated that the human activities are likely to eliminate approximately 10 million species by the year 2050. There are around 1.8 million species of plants and animals, large and microscopic, known to science in the world at present.

The number of species however is likely to be greater by a factor of at least 10. The plants and insects as well as other forms of life are not known to science are continually being identified in the worlds 'hot spots' of diversity. Unfortunately at the present rate of extinction, about 25% of the world's species will undergo extinction fairly rapidly. This may occur at the rate of 10 to 20 thousand species per year, a thousand to ten thousand times faster than the expected natural rate.

Much of this mega extinction spasm is related to human population growth, industrialization and changes in land use patterns. A major part of these extinctions will occur in 'bio-rich' areas such as tropical forests, wetlands and coral reefs. The loss of wild habitats due to rapid human population growth and the short term economic development are one of the major contributors to the rapid global destruction of biodiversity.

Habitat Loss

It is the most serious threat to wildlife. It is due to Environmental pollution.

Habitat loss also results from man's introduction of species from one area into another, disturbing the balance in existing communities. Loss of species occurs due to destruction of natural ecosystems, either for conversion to agriculture or industry or by over extraction of their resources or through pollution of air, water and soil.

Effects of Habitat Loss on Biodiversity

Habitat loss is a process of environmental change in which a natural habitat is rendered functionally unable to support the species present. It may be natural or unnatural and may be caused by habitat fragmentation, geological processes, climate change or human activities such as the introduction of invasive species or ecosystem nutrient depletion. In the process of habitat destruction, organisms that previously used the site are displaced or destroyed, reducing biodiversity.

Human destruction of habitats has accelerated greatly in the later half of twentieth century. Natural habitats are often destroyed through human activity for the purpose of harvesting natural resources for industry production and urbanization. Clearing habitats for agriculture, for example, is the principal cause of habitat destruction. Other important causes of habitat destruction include logging, mining and urban sprawl. Habitat destruction is currently ranked as the primary cause of species extinction worldwide.

Consider the exceptional biodiversity of Sumatra. It is home to one sub-species of orangutan, a species of critically endangered elephant and the Sumatran tiger, however half of Sumatra's forest is now gone. The neighboring island of Borneo, home to the other sub species of orangutan, has lost a similar area of forest and forest loss continues in protected areas.

The orangutan in Borneo is listed as endangered by the International Union for Conservation of Nature (IUCN), but it is simply the most visible thousands of species which does not survive on the disappearance of the forests of Borneo. The forests are being removed for their timber and to clear the space for plantations of palm oil, an oil used in Europe for many items including food products, cosmetics and bio-diesel.

Man-Wildlife Conflicts

Wildlife protecting bodies and local people has undergone indirect conflicts. Local communities have been affected due to loss of wildlife resources and depletion of income.

Depletion of Income

Income has depleted of rural population due to check on hunting, harvesting and human conflict is an increasing example. The elephants being cute and lovable are the cherish able animals. But if they are put next to our house they turbo violent leading to conflict with human.

Biodiversity of an area influences every aspect of lives of the people who inhabit it. Their livelihood and their living space depend on the type of ecosystem. Even people living in urban areas are dependent on the ecological services provided by the wilderness in the PAs. It is linked with the service that nature provides us. The quality of water we drink and use, the air we breathe, soil on which our food grows are all influenced by a wide variety of living organisms both plants and animals and the ecosystem of which each species is linked with in nature.

While it is well known that plant life removes the carbon dioxide and releases oxygen we breathe, it is less obvious that fungi, small soil invertebrates and even microbes are essential for plants to grow. A natural forest maintains the water in the river after the monsoon or the absence of ants could destroy life on earth, are to be appreciated to understand how we are completely dependent on the living 'web of life' on earth. This includes mankind as well. Think about this and we cannot but want to protect out earth's unique biodiversity. We are highly dependent on these living resources.

3.4.1 Endangered and Endemic Species of India

According to International Union of Conservation of Nature and Natural Resources (IUCN), the species are classified into the following types:

- Extinct Species: A species is declared as extinct, when it is no longer found in the world.

- Endangered Species: A species is declared as endangered, when its number has been reduced to a critical level. Unless it is under protection and conserved, it is in immediate danger of extinction.

- Vulnerable Species: A species is said to be vulnerable when its population is facing continuous decline due to habitat destruction of over exploitation. Such a species is still abundant.

- Rare Species: A species is said to be rare, when it is localized within restricted area or they are thinly scattered over a more extensive area. Such species are not endangered or vulnerable.

Endangered species.

Endemic Species of India

Statistics:

Category	Species Enlisted	Highly Endangered Species
Higher plants	15,000	135
Mammals	372	69
Birds	1,175	40
Reptiles & Amphibians	580	22

Fish	1,693	N.A

Equally disturbing and the matter of far more consequence is that we have till now not explored even 10% of the existing biodiversity and with such an alarming rate of extinction, we have even been destroying species, the potential of which is either not studied or less studied. In other words, the species are being destroyed with being even discovered or classified.

The causes for loss of species are complex and varied and prominent among these could be listed as follows:

- Modification, degradation and loss of habitats due to colonization and clearing of forest areas for settlement or agricultural expansion, commercial lodgings, large hydel schemes, fire, human and livestock pressure etc.

- Over exploitation, mainly for commercial (and often illegal) purposes like meat, fur, hides, body organs, medicinal etc.

- Accidental or deliberate introduction of exotic species which can threaten native flora and fauna directly by predation or by competition and also indirectly by altering the natural habitat or introducing diseases.

- Pollution (both air and water) stresses ecosystem, mismanagement of industrial and agriculture wastes threaten both terrestrial and aquatic ecosystem.

- Increase in the global surface temperature by 2° C to 6° C (global warming).

- The other possible reasons for loss of species could be improper use of agro-chemicals and pesticides, a rapidly growing human population, inequitable land distribution, economic and political policies and constraints.

3.5   Conservation of Biodiversity

Conservation is the protection, preservation, management or restoration of wildlife and natural resources such as forests and water. Through the conservation of biodiversity the survival of many species and habitats which are threatened due to human activities can be ensured.

Other reasons for conserving biodiversity include securing valuable natural resources for future generations and protecting the well being of ecosystem functions.

In-Conservation

In-conservation involves protection of fauna and flora within its natural habitat, where the species normally occurs is called in-conservation.

The natural habitats or ecosystems maintained under in-conservation are called as "protected areas".

Important In-Conservation:

Biosphere reserves, national parks, wildlife sanctuaries, Gene sanctuary, etc.

Methods of In-Conservation:

Around 4% of the total geographical area of the country is used for in-conservation. The following methods are presently used for in-conservation.

In-conservation	No. available
Biosphere reserves	7
National parks	80
Wild-life sanctuaries	420

Biosphere Reserves

Biosphere reverses cover large area, more than 5000 sq. km. It is used to protect species for long time.

Some important Biospheres Reserves in India.

Table: Biospheres Reserves in India.

S. No.	Name	Area of Bio- sphere (sq km)	Date of estab- lishment	District	State
1	Agasthyamalai	1701.00	2001	Kollam, Thiruvanan- thapuram	Kerala
2	Achanak- mar-Amarkan- tak	3835.51	2005	Anuppur, Dindori & Bilaspur	Madhya Pradesh & Chhattisgarh
3	Dibru-Saikhowa	765.00	1997	Dibrugarh and Tin- sukia	Assam
4	Dehang-Debang	5111.5	1998	Siang & Debang Valley	Arunachal Pradesh
5	Gulf of Mannar	10,500.00	1989	Indian part of Gulf of Mannar	Tamil Nadu
6	Great Nicobar	885.00	1989	Southernmost Island of Andaman and Nicobar	Andaman and Nicobar

7	Manas	2837.00	1989	Part of Kokrajhar, Bongaigaon, Barpeta, Nalbari, Kamrup and Darang	Assam
8	Kanchanjunga	2619.92	2000	Kanchanjunga Hills	Sikkim
9	Nilgiri	5520.4	1986	Part of Wayanad, Bandipur and Nagar-hole, Nilambur, Silent Valley and Siruvani Hills	Tamil Nadu, Kerala and Karnataka
10	Nanda Devi	5860	1988	Chamoli, Almora and Pithoragarh	Uttaranchal
11	Pachmarhi	4926.00	1999	Betul, Hoshangabad and Chhindwara	Madhya Pradesh
12	Nokerek	80.00	1988	Part of Garo Hills	Meghalaya
13	Sunderbans	9630.00	1989	Delta of Ganges and Brahmaputra	West Bengal
14	Similipal	4374.00	1994	Mayurbhanj	Orissa

Role of Biosphere reserves:

- It gives long-term survival of evolving ecosystem.

- It protects endangered species.

- It serves as site of recreation and tourism.

- It is also useful for educational and research purposes.

Restriction:

- No tourism and explosive activities are permitted in the biosphere reserves.

National Park

A National Park is an area dedicated for the conservation of wildlife along with its environment. It is usually a small reserve covering an area of about 100 to 500 sq.ft. Within the biosphere reserves, one or more national parks are also exists.

Role of a National Park:

- It is used for enjoyment through tourism, without affecting the environment.

- It is used to protect, propagate and develop the wildlife.

Restrictions:

- Grazing of domestic animals inside the national park is prohibited.

- All private rights and forestry activities are prohibited within a national park.

Wildlife Sanctuaries

A wildlife sanctuary is an area, which is reserved for the conservation of animals only. At present, there are 492 wildlife sanctuaries in our country.

Role of Wildlife Sanctuaries:

- It allows the operations such as collection of forests products, harvesting of timer, private ownership rights and forestry operations provided.
- It does not affect the animals adversely.
- It protects animals only.

Restrictions:

- Killing, hunting, shooting or capturing of wildlife is prohibited except under the control of higher authority.

Gene Sanctuary

A gene sanctuary is an area, where the plants are conserved.

Example: In Northern India, two gene sanctuaries are found available:

- One gene sanctuary for citrus.
- One gene sanctuary for pitcher plant.

Advantages of In-Conservation:

- It is cheap and convenient method.
- The species gets adjusted to the natural disasters like drought, floods and forest fires.

Disadvantages:

- Maintenance of the habitats is not proper, due to shortage of staff and pollution.
- Large surface area of the earth is required to preserve the biodiversity.

Ex-Conservation

Ex- conservation involves protection of fauna and flora outside the natural habitats.

This type of conservation is mainly done for conservation of crop varieties and the wild relatives of crops.

Role of Ex-Conservation:

- It identifies those species which are at more risk of extinction.

- It involves maintenance and breeding of the endangered plant and animal species under controlled conditions.

- It prefers the species, which are more important to man in near future among the endangered species.

Important Ex-Conservation:

Botanical gardens, seed banks, microbial culture collections, tissue and cell cultures, museums.

Methods of Ex- Conservation:

The following important gene bank or seed bank facilities are used in Ex-conservation:

National Bureau of Plant Genetic Resources (NBPGR)

It is located in New Delhi. It uses preservation techniques to preserve agricultural and horticultural crops.

Preservation Technique:

It involves the preservation of seeds, pollen of some important agricultural and horticultural crops by using liquid nitrogen at a temperature as low as - 196°C. Varieties of rice, turnip, radish, onion, have been preserved successfully in liquid nitrogen for several years.

National Bureau of Animal Genetic Resources (NBAGR)

It preserves the semen of domesticated bovine animals.

National Facility for Plant Tissue Culture Repository (NFPTCR)

It develops the facility for conservation of varieties of crop plants or tree by tissue culture. This facility has been created within the NBPGR.

Advantages of Ex-Conservation:

- Survival of endangered species is increasing due to special care and attention.

- It captive breeding, animals are assured food, water, shelter and also security and hence longer life Span.

- It is carried out in cases of the endangered species, which do not have any chances of survival, in the wild.

Disadvantages:

- The animals cannot survive on natural environment.

- It is expensive method.

- The freedom of wildlife is lost.

Chapter 4

Environmental Pollution

4.1 Definition, Cause, Effects and Control Measures of Environmental Pollutions

Pollution can be defined as the presence of a substance which has harmful or poisonous effects.

Causes

Pollution occurs in different forms such as Air, water, soil, radioactive, noise, heat/ thermal and light. Every form of the pollution has two sources of occurrence. The point and the non-point sources. The point sources are easy to identify, monitor and control, whereas the nonpoint sources are hard to control.

4.1.1 Air Pollution

The presence of one or more contaminants like smoke, dust, mist and odor in the atmosphere which are injurious to human beings, plants and animals.

Common Air Pollutants Sources and their Effects

According to the World Health Organization (WHO), more than 1.1 billion people live in urban areas where outdoor air is unhealthy to breath. Some of the common air pollutants are as follows:

Carbon monoxide (CO)

It is a colorless gas that is poisonous to air breathing animals. It is usually formed during the incomplete combustion of carbon containing fuels.

$$2C + O_2 \rightarrow 2CO$$

Human Source: Cigarette smoking, incomplete burning of fossil fuels. About 77% comes from motor vehicle exhaust.

Health Effects

Reacts with hemoglobin in red blood cells and reduces the ability to bring oxygen to body cells and tissues by the blood. Which can cause anemia and headaches. At high levels it leads to coma, irreversible brain cell damage and death.

Nitrogen dioxide (NO_2)

It is a reddish-brown irritating gas that gives the photochemical smog. In the atmosphere it can be converted into nitric acid.

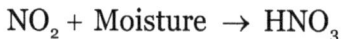

$$NO_2 + Moisture \rightarrow HNO_3$$

Human Source: Fossil fuel burning in motor vehicles (49%) and power industrial plants (49%).

Health Effects: Lungs irritation and damage.

Environmental Effects: Acid deposition of HNO^3 can damage trees, soils and aquatic life in lakes, HNO^3 can corrode metals and eat away stone on buildings, statues and monuments. NO^2 can damage fabrics.

Sulfur dioxide (SO_2)

It is a colorless and irritating gas. It is formed mostly from the combustion of sulfur containing fossil fuels such as coal and oil. In the atmosphere it can be converted to sulfuric acid which is a major component of acid deposition.

Human Sources: Coal burning in power plants (88%) and industrial processes (10%).

Health Effects:

Breathing problems for healthy people.

Environmental Effects: Reduce visibility, acid deposition of H^2SO^4 can damage trees, soils and aquatic life in lakes.

Hydrocarbons

Hydrocarbons especially lower hydrocarbons gets accumulator due to the decay of vegetable matter.

Human Source: Agriculture, decay of plants and burning of wet logs.

Health Effects: Carcinogenic.

Ozone (O$_3$)

Highly reactive irritating gas with an unpleasant odor that forms in the troposphere. It is a major component of the photochemical smog.

Human Source: Chemical reaction with volatile organic compounds and nitrogen oxides.

Environmental Effects: Moderates the climate.

Chromium (Cr)

It is solid toxic metal, emitted into the atmosphere as particulate matter.

Human Source: Paint, Smelters, Chromium plating.

Health Effects: Perforation of nasal septum, chrome holes.

Control Measures

The atmosphere has several built-in self-cleaning processes such as dispersion, gravitational settling, flocculation, absorption, rain washout and so on, to cleanse the atmosphere.

In terms of a long range of air pollution, control of contaminants at their source is a more desirable and effective method through preventive or control or technologies.

Source Control:

Since we know the substances that cause air pollution, the first approach to its control will be through source reduction.

- Use only unleaded petrol.

- Use petroleum products and other fuels that have low sulfur and ash content.

- Reduces the number of private vehicles on the road by developing an effective public transport system.

- Ensure that houses, schools, restaurants and places where children play are not located on busy streets.

- Plant trees along busy streets because they remove particulates and carbon monoxide and absorb noise.

- Industries and waste disposal sites should be situated outside the city center preferably downwind of the city.

- Use catalytic converters to help control the emission of carbon monoxide and hydrocarbons.

Control measures in industrial centers:

- The emission rates should be restricted to permissible levels by each and every industry.

- Continuous monitoring of the atmosphere for the pollutants should be carried out to know the emission levels.

- Incorporation of air pollution control equipment's in the design of the plant layout must be made mandatory.

Public Health Aspects

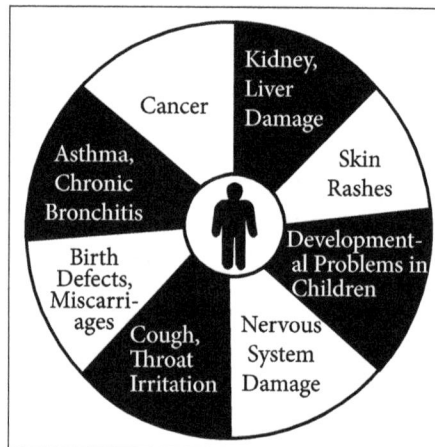

Effects on public health.

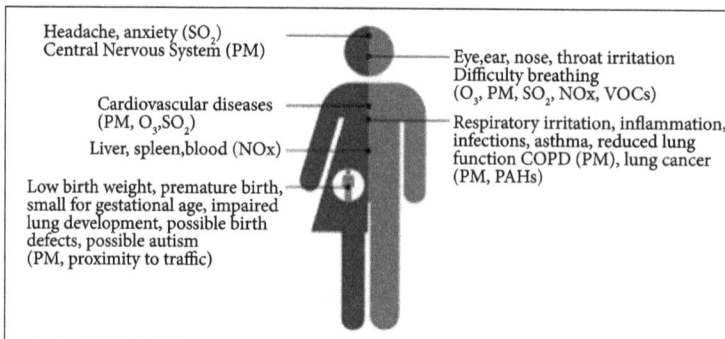

The Air Pollution and Respiratory Health Branch lead CDC's fight against the environmental related respiratory illnesses, including asthma and studies indoor and outdoor air pollution.

Air Pollution and Repository health:

- Asthma: It is a serious environmental health threat, but it can be controlled by taking the medication and by avoiding contact with the environmental "triggers" such as furry pets, dust mites, tobacco smoke, mold and certain chemicals.

- Mold: Exposure to damp and moldy environments may cause throat irritation, nasal stuffiness, eye irritation, coughing or wheezing or skin irritation.

- Carbon Monoxide Poisoning: Carbon monoxide (CO), which is an odorless, colorless gas that can cause sudden illness and death, is found in combustion fumes produced by generators, cars and trucks, lanterns, stoves, gas ranges, burning charcoal and wood and heating systems.

Climate and Health

Climate and Health Program works to prevent and adapt to the health impacts of the extreme weather and the other climate related issues.

Environmental Public Health Tracking

Environmental Public Health Tracking is the ongoing collection, the integration, analysis, interpretation and dissemination of data on environmental hazards, exposures to those hazards and health effects which may be related to the exposures. The goal of tracking is to provide information which can be used to plan, apply and evaluate actions to prevent and control environmentally related diseases.

Health Studies

The Health Studies Branch investigates the human health effects of exposure to environmental hazards ranging from chemical pollutants to natural, technologic or terrorist disasters. The results are used to develop, implement and evaluate strategies which can help in preventing or reducing harmful exposures.

Radiation Studies

The Radiation Studies Branch identifies the potentially harmful environmental exposures to ionizing radiation and associated toxicants, conducts the energy related health research and responds to protect public's health in the event of an emergency involving radiation or radioactive materials.

4.1.2 Water Pollution

Water pollution is the pollution caused due to any chemical, physical or biological change in water quality that has a harmful effect on living organisms or makes water unsuitable for desired uses.

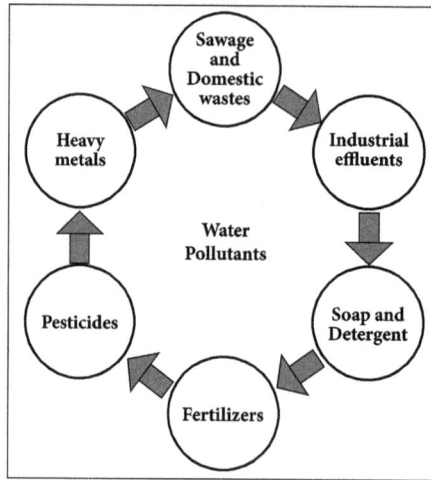

Water pollutants.

Infection Agents

Example: Bacteria, Viruses, Protozoa.

Human Sources: Human and animal wastes.

Effects: Variety of diseases.

Oxygen Demanding Wastes

Example: Organic wastes such as animal manure and plant debris that can be decomposed by aerobic bacteria.

This degradation consumes dissolved oxygen in water. DO is the amount of oxygen dissolved in a given quantity of water at a particular pressure and temperature.

The standard point of DO varies from 8 - 15 mg/lit.

Human Source: Sewage, animal feedlots, paper mills.

Effects: Large populations of bacteria decomposing these wastes can degrade water quality by depleting water of dissolved oxygen. This causes fish and other forms of oxygen consuming aquatic life to die.

Organic Chemicals

Example: Oil, Gasoline, Plastics, Pesticides.

Human Sources: Industrial effluents, surface runoff from farms, household cleansers.

Effects:

- Can threaten human health by causing nervous system damage and some cancers.

- Harm fish and wildlife.

Sediment

Examples: Soil, Silt, etc.

Human Sources: Land erosion.

Effects:

- Can cloud water and reduce photosynthesis.

- Disrupt aquatic food webs.

- Carry pesticides, bacteria and other harmful substances.

- Settle out and destroy feeding and spawning rounds of fish.

Radioactive Materials

Example: Radioactive isotopes of iodine, radon, uranium and chromium.

Human Sources: Nuclear power plants, mining and processing of uranium and other ores, nuclear weapons production and natural sources.

Effects: Genetic mutations, birth defects and certain cancers.

Control measures of water pollution:

- Administration of water pollution control should be in the hands of state or central government.

- Scientific techniques should be adopted for environmental control of catchment areas of rivers, ponds or streams.

- Industrial plants should be based on recycling operations as it helps prevent disposal of wastes into natural waters but also extraction of products from waste.

- Plants, trees and forests control pollution as they act as natural air conditioners.

- Trees are capable of reducing sulphur dioxide and nitric oxide pollutants and hence more trees should be planted.

- No type of waste (treated, partially treated or untreated) should be discharged into any natural water body. Industries should develop closed loop water supply schemes and domestic sewage must be used for irrigation.

- Qualified and experienced people must be consulted from time to time for effective control of water pollution.

- Public awareness must be initiated regarding adverse effects of water pollution using the media.

- Laws, standards and practices should be established to prevent water pollution and these laws should be modified from time to time based on current requirements and technological advancements.

- Basic and applied research in public health engineering should be encouraged.

4.1.3 Soil Pollution

Soil pollution is caused due to the addition of chemicals which reduce its productive capacity. In addition to the urban solid wastes, several hazardous chemicals are dumped into the soil. The toxic substances from these dumps leach out and percolate through the soil and contaminate the ground water. In agricultural operations, most of fertilizers and pesticides are used and the residual chemicals remain in the top layers of soil. Toxic insecticides kill useful soil bacteria that are favorable for plant growth.

Soil acts like a filter in removing the impurities in water and waste waters. But toxic residual chemicals from the soil reach human beings through vegetables, fruits, etc. Industrial effluents and solid wastes without adequate treatment which get deposited on the land or in water bodies are major sources of soil pollution.

Various pollutants and their harmful effects on soil are as follows:

- Organic wastes enter the soil pores and decompose. Pathogenic bacteria present in these wastes spread infection. Hookworm and also cause diseases.

- Compounds containing arsenic, mercury, chromium, nickel, lead, cadmium, zinc and iron are toxic to life. Fluorides also affect the plant growth.

- Excess use of sodium, magnesium, calcium, potassium, sulfur, zinc and iron in the form of fertilizers and pesticides inhibit plant growth and reduce crop yield. It is necessary to keep the dosage at an optimal level.

- Water logging and salinity increases the dissolved salt content in ground water and the soil. Some plants are too much sensitive to soil pH and salinity. High salinity may make the land unfit for cultivation.

Soil erosion also causes soil pollution as given below:

- Soil erosion is the removal of soil from its original place. Water, ice wind and other climatic agents promote soil erosion. Cutting of trees, construction and other human activities reduce the vegetation cover which protects the soil from sun, rain, wind and moving water and as a result the soil is eroded.

- Due to soil erosion essential minerals are removed resulting in the loss of fertility of soil. Soil erosion results in the increased sedimentation of rivers and lakes which has an impact on water quality and subsequently on the populations of aquatic organisms including fish.

- It is not advisable to construct any new building on soils such as clay and silt that have been recently carried forward and deposited by the waters of rivers and streams as the load bearing capacity of such a soil medium is very low. There are cases where the underground stratum is peat deposited by sewage canals that had existed before. Such sites should never be selected for the construction of buildings.

- In a natural ecosystem, essential nutrients and minerals cycle from the soil to living organisms and back to the soil when the organisms die. This naturals system is disrupted by agricultural operations. In dry climates much of the water used for irrigation evaporates, leaving behind high concentration of salts, such as sodium chloride, in the top soil, the accumulation of salts in the top soil is known as salinization.

- For the conservation of soil, various methods are used to reduce soil erosion, to prevent depletion of soil nutrients and to restore nutrients already lost by erosion, leaching and excessive crop harvesting.

Control Measures of Soil Pollution

- Soil erosion can be controlled by variety of forestry and farm practices. Ex: Planting trees on barren slopes.

- Contour cultivation and strip cropping may be practiced instead of shifting cultivation.

- Terracing and building diversion channels may be undertaken.

- Reducing the deforestation and substituting chemical manures by animal wastes also helps arrest soil erosion in long term.

- Proper dumping of unwanted materials: Excess wastes by man and animals pose a disposal problem. The open dumping is the most commonly practiced technique. Nowadays, controlled tipping is followed for solid waste disposal. The surface so obtained is used for housing or sports field.

- Production of natural fertilizers: Bio-pesticides should be used in place of toxic chemical pesticides. The organic fertilizers should be used in place of synthesized chemical fertilizers. Ex: The organic wastes in animal dung may be used to prepare compost manure instead of throwing them wastefully and polluting the soil.

- Proper hygienic condition: The people should be trained regarding sanitary habits. Ex: Lavatories should be equipped with quick and effective disposal methods.

- Public awareness: Informal and formal public awareness programs should be imparted to educate people on health hazards by environmental education. Ex: Mass media, Educational institutions and voluntary agencies can achieve this.

- Recycling and Reuse of wastes: To minimize soil pollution, the wastes such as paper, plastics, metals, glasses, organics, petroleum products and industrial effluents etc. should be recycled and reused. Ex: The industrial wastes should be properly treated at source. Integrated waste treatment methods should be adopted.

- Ban on Toxic chemicals: The ban should be imposed on chemicals and pesticides like DDT, BHC, etc. which are fatal to plants and animals. The nuclear explosions and improper disposal of radioactive wastes should be banned.

4.1.4 Noise Pollution

The unwanted, unpleasant or disagreeable sound that causes discomfort for all living things.

Noise Pollution.

The sound intensity is measured in decibel (dB), which is tenth part of the longest unit Decibel, One dB is equal to the faintest sound, a human ear can hear.

Types of Noise

It has been found that the environmental noise is being doubling for every 10 years. Generally noise is described as:

- Industrial noise.

- Transport noise.

- Neighborhood noise.

Industrial Noise: Highly intense sound or noise pollution is caused by many machines. There exists a long list of sources of noise pollution including different machines of numerous factories, industries and mills.

Industrial noise, particularly from mechanical saws and pneumatic dwell is unbearable and nuisance to public.

Example: In the start industry, the workers near the heavy industrial blowers are exposed to 112 dB for eight hours and suffer from the occupational pollution.

Transport Noise: The main noise comes from transport. It mainly includes road traffic noise, rail traffic noise and air craft noise. The number of road vehicles like motors, scooters, cars and particularly the diesel engine vehicles have increased enormously in recent years.

A survey conducted in metropolitan cities has shown that noise level on Delhi, Bombay and Calcutta is as high as 90 dB. Inhabitants of cities are subjected to the most annoying form of transport noise which gradually deafens them.

Neighborhood Noise: This type of noise includes disturbances from the household gadgets and community. Common noise makes are musical instruments, TV, VCR, radios, transistors, telephones and loudspeakers, etc. Ever since the industrial revolution, noise in environment has been doubling every ten years.

Effects of Noise Pollution

- It causes the muscles to contract leading to nervous breakdown, tension, etc.

- Noise pollution affects human comfort, health and efficiency. It causes contraction of blood vessels, makes the skin pale and leads to excessive secretion of the adrenalin hormone into blood stream which is responsible for high blood pressure.

- It affects health efficiency and behavior. It may cause damage to heart, brain, kidneys and lives and may also produce emotional disturbances.

- The adverse reactions are coupled with a change in hormone content of blood, which in turn increase the rate of heart beat, construction of blood vessels, digestive spasms and dilation of pupil of eye.

- Recently it has been reported that blood is also thickened by excessive noises.
- In addition to the serious loss of hearing due to excessive noise, impulsive noise also causes psychological and pathological disorders.

Control Measures

Source control: This may include source modification such as acoustic treatment to machine surface, design changes, limiting the operational timings and so on.

Transmission path intervention: This may include containing the source inside a sound insulating enclosure, construction of noise barrier or provision of sound absorbing materials along the path.

Receptor control: This includes protection of the receives by altering the work schedule or provision of personal protection devices such as ear plugs for operating noisy machinery. The measure may include dissipation and deflection methods.

Oiling: Proper oiling will reduce the noise from the machines.

Preventive Measures

Noise can be reduced by prescribing noise limits for vehicular traffic, ban on honking of horns in certain areas and creation of silent zones near the schools and hospitals and redesigning of buildings to make them noise proof. Other measures can involve reduction of traffic density in residential areas and giving preferences to the mass public transport system.

4.1.5 Nuclear Hazards

It is defined as the risk or danger to human health or environment posed by radiation emanating from atomic nuclei of a given substance or the possibility of an uncontrolled explosion originating from a fusion or fission reaction of atomic nuclear.

Sources of Nuclear Hazards: Sources of radioactivity are both natural and manmade.

Natural Sources Include

1. Cosmic rays from outer space The quantity depends on the altitude and latitude. It is more at higher latitudes and high altitudes.

2. Emissions from radioactive materials from Earth's crust.

People have been exposed to low levels of radiation from these natural sources for several. But it is manmade sources which are posing a threat to mankind. The man-made sources of radioactivity are nuclear wastes produced at the time of:

- Uses of radioactive material in nuclear power plants.

- Mining and processing of radioactive ores.

- Uses of radioactive materials in nuclear weapons.

- Uses of radioactive isotopes in medical, industrial and research applications.

The greatest exposure to human beings comes from the diagnostic use of the X-rays, radioactive isotopes used as the tracers and the treatment of cancer and other ailments.

Effects of Nuclear Hazards

Effects of radioactive pollutants depend upon half-life, energy releasing capacity, rate of diffusion and rate of deposition of the contaminant. Various atmospheric conditions and climatic conditions such as temperature, wind and rainfall also determine their effect. All organisms are affected from the radiation pollution and the effects are extremely dangerous. The effects may be somatic or genetic damage. The effects are cancer, shortening of life span and genetic effects or mutations. Some of the possible effects are listed as below:

- Exposure at low doses of radiations (100-250 rads), men do not die but begin to suffer from fatigue, nausea, vomiting and loss of hair. But recovery can be possible.

- Radiations may break the chemical bonds such as DNA in cells. This affects the genetic makeup and control mechanisms. The effects can be instantaneous, prolonged or delayed types. Even it could be carried to future generations.

- Higher irradiation doses (10,000rads) kills the organisms by damaging the tissues of heart, brain, etc.

- Exposure at higher doses (400-500 rads), blood cells are reduced, bone marrow is affected, blood fails to clot and the irradiated person soon dies of infection and bleeding.

- Through food chain also, the radioactivity effects are experienced by man. But the most significant effect of radioactivity causes long range effects, natural resistance and fighting capacity against germs is reduced, affecting the future of man and hence the future of our civilization.

- Workers handling the radioactive wastes get slow but continuous irradiation and in course of time develop cancer of different types.

Control of Nuclear Hazards

The peaceful uses of radioactive materials are so wide and effective that modern civilization cannot go without them on the other hand; there is no cure for radiation damage. Hence, the only option against nuclear hazards is to check and prevent radioactive pollution:

- Safety measures are enforced strictly.

- Leakages from nuclear reactors, careless handling, transport and use of radioactive fuels, fission products and radioactive isotopes have to be totally stopped.

- There should be regular monitoring and quantitative analysis through frequent sampling in the risk areas.

- Waste disposal is careful, efficient and effective.

- Appropriate steps should be taken against occupational exposure.

- Preventive measures should be followed so that background radiation levels do not exceed the permissible limits.

- Safety measures should be strengthened against the nuclear accidents.

Disposal of Nuclear Wastes

Since nuclear waste can be extremely dangerous and therefore, the way in which they are to be disposed of is strictly controlled by international agreement. Since 1983, by international agreement, the disposal in the Atlantic Ocean and into the atmosphere has been banned. After processing, to recover usable material and reducing the radioactivity of the waste, disposal is made in solid form where possible. The nuclear wastes are classified into three categories such as:

High Level Wastes (HLW): It has a very high radioactivity per unit volume. For example, spent nuclear fuel. Have to be cooled and are, therefore, stored for several decades by its producer before disposal. Since these wastes are too dangerous to be released anywhere in the biosphere, they must be contained either by converting them into inert solids and then buried deep into earth or are stored in deep salt mines.

Medium level wastes (MLW): They are solidified and are mixed with concrete in steel drums before being buried in deep mines or below the sea bed in concrete chambers.

Low liquid wastes (LLW): They are disposed of in steel drums in concrete lined trenches in designated sites.

4.2   Role of an Individual in Prevention of Pollution

Environmental pollution cannot be prevented and removed. The proper implementation and particularly the individual participation are the important aspects which should be given due importance.

The individual participation is useful in law making processes and restraining the pollution activities and thereby the public participation plays a major role in the effective environmental management. A small effect is made by each individual at his own place will have pronounced effect at the global level. It is suitably said "Think globally act locally."

Each individual should change his or her life style in such a way as to reduce environmental pollution.

Individual Participation

- Help more in pollution prevention than pollution control.

- Plant more trees.

- Purchase recyclable, recycled and environmentally safe products.

- Use water, energy and other resources efficiently.

- Use natural gas than coal.

- Use CFC free refrigerators.

- Increase use of renewable resources.

- Reduce deforestation.

- Use office machines in well ventilated areas.

- Remove Nox from motor vehicular exhaust.

4.2.1 Pollution Case Studies

Case Study: The Ganga, India

There is a universal reverence to water in almost all of the major religions of the world. Most religious beliefs involve some ceremonial use of "holy" water. The belief in its known historical and unknown mythological origins, the purity of such water and the inaccessibility of remote sources, elevate its importance even further.

In India, the river Ganga occupies a unique position in the cultural ethos of India. Legend says that the river has descended from the Heaven on earth as a result of the long and arduous prayers of King Bhagirathi for the salvation of his deceased ancestors.

From times immemorial, the Ganga has been India's river of faith, devotion and worship. Millions of Hindus accept its water as sacred. Even today, people carry the treasured Ganga water all over India and abroad because it is a "holy" water and known for its "curative" properties.

Ganga River

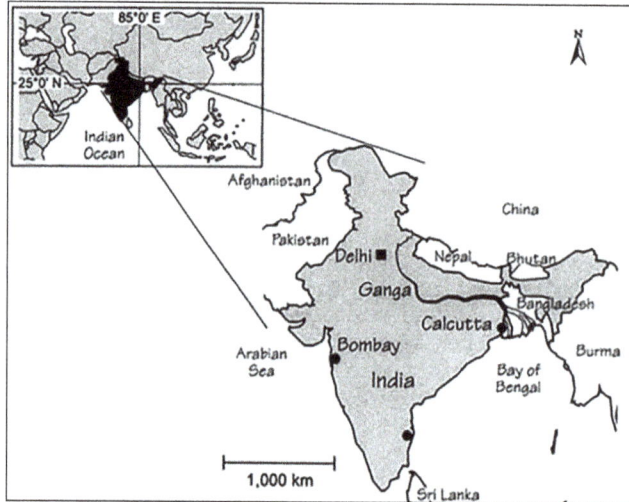

Location map of India showing the Ganga River.

The Ganga rises on the southern slopes of the Himalayan ranges from the Gangotri glacier at 4,000m above mean sea level. It flows swiftly for 250km in the mountains, descending steeply to an elevation of 288m above mean sea level. In the Himalayan region, Bhagirathi is joined by the tributaries Mandakini and Alaknanda to form the river Ganga. After entering the plains at Haridwar, it winds its way to the Bay of Bengal, covering 2,500km through the provinces of Bihar, Uttar Pradesh and West Bengal. In the plains, it is joined by Ramganga, Yamuna, Sai, Gomti, Ghaghara, Sone, Gandak, Kosi and Damodar along with many other smaller rivers.

Map of India showing the route of the Ganga River.

The purity of the water depends on the velocity and the dilution capacity of the river. A large part of the flow of the Ganga is abstracted for irrigation just as it enters the plains at Hardiwar. The Ganga receives over 60 per cent of its discharge from its tributaries. The contribution of most of the tributaries to the pollution load is small, except from the Damador, Gomti and Yamuna rivers, for which separate action programs were already started under Phase II of "The National Rivers Conservation Plan".

Exploitation

In the recent past, due to rapid progress in communications and commerce, there has been a swift increase in the urban areas along the river Ganga, As a result the river is no longer only a source of water but is also a channel, receiving and transporting urban wastes away from the towns. Today, 1/3rd of the country's urban population lives in the towns of the Ganga basin. Out of the 2,300 towns in the country, 692 are located in this basin and of these, 100 are located along the river bank itself.

The belief the Ganga river is "holy" has not, however, prevented over-use, abuse and pollution of the river. Due to over-abstraction of water for irrigation in the upper regions of the river, the dry weather flow has been reduced to a trickle. Rampant deforestation in the last few decade's results in topsoil erosion in the catchment area, has increased silt deposits which, in turn, raise the river bed and lead to devastating floods in the rainy season and stagnant flow in the dry season. Along the main river course, there are 25 towns with a population of more than 100,000 and about another 23 towns with populations above 50,000. In addition there are 50 smaller towns with populations above 20,000. The natural assimilative capacity of the river is severely stressed.

Sources of Pollution of the Ganga

The principal sources of pollution of the Ganga River can be characterized as follows:

- Solid garbage thrown directly into the river.

- Domestic and industrial wastes: It has been estimated that about $1.4 \times 10_6 m_3 d_{-1}$ of the domestic wastewater and $0.26 \times 10_6 m_3 d_{-1}$ of the industrial sewage are going into the river.

- Animal carcasses and half-burned and unburned human corpses are thrown into the river.

- Non-point sources of pollution from agricultural run-off containing residues of harmful pesticides and fertilizers.

- Mass bathing and ritualistic practices.

- Defecation on the banks by low-income people.

Ganga Action Plan

Scientific Awareness

There are 14 major river basins in India with natural waters which are being used for human and developmental activities. These activities contribute to the pollution loads of these river basins. Of these river basins, the Ganga sustains the largest population. The Central Pollution Control Board (CPCB), which is India's national body for monitoring the environmental pollution, undertook a comprehensive scientific survey in 1981-82 to classify the river waters according to their designated best uses. This report was the first systematic document that formed the basis of the Ganga Action Plan (GAP). It detailed the land-use patterns, domestic and industrial pollution loads, fertilizer and pesticide use, hydrological aspects and river classifications. This inventory of pollution was used by the Department of Environment in the year 1984 when formulating a policy document.

Realizing the need for urgent intervention of the Central Ganga Authority was set up in 1985 under the chairmanship of the Prime Minister. The Ganga Project Directorate was established in June 1985 as a national body operating within the National Ministry of Environment and Forest. The GPD serve as the secretariat to the CGA and also as the Apex Nodal Agency for implementation. It was set up to co-ordinate the different ministries involved and to administer funds for this 100 per cent centrally-sponsored plan.

The programme was perceived as a once-off investment providing demonstrable effects on the river water quality. The execution of works and its subsequent operation and management (O&M) were the responsibility of the state governments, under the supervision of GPD. GPD was to remain in place until the GAP was completed. The plan was formally launched on 14 June 1986. The main thrust was to divert and intercept the wastes from urban settlements away from the river. Treatment and economical uses of waste were made an integral part of the plan. It was realized that the comprehensive co-ordinated research would have to be conducted on the following aspects of Ganga:

- A more rational plan for the use of the resources of the Ganga for animal husbandry, agriculture, forests, fisheries, etc.

- The sources and nature of the pollution.

- The possible revival of the inland water transport facilities of the Ganga together with the tributaries and distributaries.

- The demographic, cultural and human settlements on the banks of the river.

One outcome of this initiative was a multidisciplinary study of the river in which 14 universities located in the basin participated in a well co-ordinated, integrated research programme. This was one of the largest endeavors, involving several hundred scientists, ever undertaken in the country and was funded under GAP. The resultant report is a unique, integrated profile of the river.

GAP was the first step in river water quality management. Its mandate was limited to quick and effective, but sustainable, interventions to contain the damage. The studies carried out by the CPCB in 1981-82 revealed that pollution of the Ganga was increasing but had not assumed serious proportions, except at the certain main towns on the river such as industrial Kanpur and Calcutta on the Hoogly, together with a few other towns. The causative factors responsible for these situations were targeted for swift and effective control measures. This strategy was adopted for urgent implementation during the first phase of the plan under which only 25 towns identified on the main river were to be included.

The studies had revealed that:

- 8 percent of the municipal sewage was from the 25 Class I towns on the main river.

- 75 percent of the pollution load was from untreated municipal sewage.

- All the industries accounted for only 25 percent of the total pollution.

- Only a few of these cities had sewage treatment facilities.

Attainable Objectives

The broad aim of the GAP was to reduce the pollution, clean the river and to restore the water quality at least to Class B (i.e. bathing quality: $3 \, mg \, l_{-1}$ BOD and $5 \, mg \, l_{-1}$ dissolved oxygen). This was considered as a feasible objective and because a unique and distinguishing feature of Ganga was its widespread use for ritualistic mass bathing.

The multipronged objectives were to improve the water quality by controlling the municipal and industrial wastes. The long-term objectives were to improve the environmental conditions along the river by suitably reducing all the polluting influences at source.

Prior to the creation of the GAP, responsibilities for pollution of the river were not clearly demarcated between the various government agencies. The pollutants reaching Ganga from most point sources did not mix well in the river, due to the sluggish water currents and as a result such pollution often lingered along the embankments where people bathed and took water for domestic use.

Strategy

The GAP had a multipronged strategy to improve the river water quality. It was fully financed by central Government, with the assets created by the central Government to be used and maintained by the state governments. The main thrust of the plan was targeted to control all municipal and industrial wastes.

All the possible point and non-point sources of pollution were identified. The control of urban non-point sources was also tackled by direct interventions from the project funds. The control of point sources of urban municipal wastes for the 25 Class I towns on the main river was initiated from 100percent centrally-invested project funds. The control of non-point source agricultural run-off was undertaken in a phased manner by the Ministry of Agriculture, principally by reducing the use of fertilizer and pesticides. The control of point sources of industrial wastes was done by applying the polluter-pays-principle.

A total of 261 sub-projects were sought for implementation in 25 Class I river front towns. This would eventually involve a financial outlay of Rs.4,680 million (Indian Rupees), equivalent to about US \$156 million. More than 95 percent of the programme has been completed and remaining sub-projects are in various stages of completion. The resultant improvement in the river water quality is hotly debated in the media by certain non-governmental organizations (NGOs). The success of the programme can be gauged by fact that Phase II of the plan, covering some of the tributaries, has already been launched by the Government. In addition, the earlier action plan has now evolved further to cover all the other major national river-basins in India, including a few lakes and is known as the "National Rivers Conservation Plan".

Domestic Waste

The major problem of pollution from the domestic municipal sewage ($1.34 \times 10_6 \, m_3 d_{-1}$) arising from the 25 selected towns were handled directly by financing the creation of facilities for diversion, interception and treatment of wastewater and also by preventing the other city wastes from entering the river. Out of the $1.34 \times 10_6 m_3 d_{-1}$ of sewage assessed to be generated, $0.873 \times 10_6 m_3 d_{-1}$ was intercepted by laying 370 km of trunk sewers with 129 pumping stations as part of 88 sub-projects.

The activities of various sub-projects can be summarized as follows:

Approach to river water quality improvement	Number of schemes
Sewage treatment plants	35
Interception and diversion of municipal wastewater	88
Electric crematoriums	28
Low-cost sanitation complexes	43
River front facilities for bathing	35
Others (e.g. biological conservation of aquatic species, river quality monitoring)	32
Total	261

A total of 248 of these schemes have already been commissioned and the remaining is due to be completed by 1998.

Industrial Waste

About 100 industries were identified on the main river itself. Sixty-eight of these were considered grossly polluting and were discharging $260 \times 10_3$ m_3d_{-1} of wastewater into the river. Under the Water Act 1974 and Environment Act 1986, 55 industrial units (generating $232 \times 10_3 m_3 d_{-1}$) out of the total of 68 grossly polluting industrial units complied and installed effluent treatment plants. In addition, two others have treatment plants under construction and currently one unit does not have a treatment plant. Legal proceedings have been taken against the remaining 12 industrial units which were closed down for non-compliance.

Applied Research

The Action Plan stressed the importance of applied research projects and many universities and reputable organizations were supported with grants for projects carrying out studies and observations which would have a direct bearing on the Action Plan.

Some of the prominent subjects were sewage-fed pisciculture, PC-based software modeling, bio conservation in Bihar, conservation of fish in upper river reaches, using treated sewage for irrigation, monitoring of pesticides and rehabilitation of turtles.

Some of the ongoing research projects include land application of untreated sewage for tree plantations, disinfection of treated sewage by UV radiation, aquaculture for sewage treatment and disinfection of treated sewage by Gamma radiation. All the presently available research results are being consolidated for easy access by creation of a data base by the Indian National Scientific Documentation Centre.

Integrated Improvements of Urban Environments

The GAP also covered very wide and diverse activities, such as conservation of aquatic species (gangetic dolphin), protection of natural habitats and creating riverine sanctuaries. It also included components for building stepped terraces on the sloped river banks for ritualistic mass-bathing (128 locations), landscaping river frontage (35 schemes), development of public facilities, improving sanitation along the river frontage (2,760 complexes), improved approach roads and lighting on the river frontage.

Public Participation

The pollution of the river, although classified as environmental, was the direct outcome of the deeper social problem emerging from long-term public diffidence and apathy, indifference and lack of public awareness, education and social values and above all from poverty.

In recognition of necessity of the involvement of the people for sustainability and success of the Action Plan, importance was given for generating awareness through intensive publicity campaigns using the audio visual approaches, press and electronic media, leaflets and hoardings as well as organizing public programs for spreading the message effectively. In spite of full financial support from the project and heavy involvement of about 39 well known NGOs to organize these activities, the programme had only limited public impact and even received some criticism. Other similar awareness-generating programs involving school children from many schools in the project towns were received with greater enthusiasm.

Operation and Maintenance

The enduring success of the pollution abatement works under the GAP is essential for sustainability. Most of these works were carried out by the same agencies which were eventually responsible for maintaining them as part of their primary functions, such as the municipality, the city development authority or the irrigation and flood control department. The responsibility for subsequent O&M of these works automatically passed to these agencies.

The most crucial components for preventing the river pollution were the main pumping stations which were intercepting the sewage and diverting it to the treatment plants. These large capacity pumping stations, operating at the city level, had been built for the first time in India and it was considered that the municipalities would have adequate skilled personnel and resources to be able to manage them.

The central Government shared half of this deficit until 1997. In the broader interest of pollution control, future policies will also be similar, where the state governments undertake the responsibility for pollution control works because the local bodies are unable to bear the cost of O&M expenditures with such limited resources.

Technology Options

The choice of technology for the GAP was largely conventional based on the available options and local considerations. Consequently, the sewers and pumping stations and all the similar municipal and conservancy works were executed in each province by its own implementing agencies, according to their customary practices but within the commonly prescribed specifications. The choice of technology for most of the large domestic wastewater treatment plants was carefully decided by a panel of experts in close consultation with those external aid agencies. A parallel procedure was adopted in-house for all other similar projects. For all the larger sewage treatment plants, the unanimous choice was to adopt the well-accepted activated sludge process. Some of these new and simpler technologies, with their low-cost advantages, will emerge as the large-scale future solution to India's sanitation problems.

Implementation Problems

The implementation of a project of this magnitude over the entire 2,500 km stretch of the river, covering 25 towns and crossing three different provinces, which could be achieved by delegating the actual implementation to the state government agencies which had the appropriate capabilities. The state governments also undertook the responsibility of subsequently operating the assets being created under the programme and maintaining them.

The overall inter-agency co-ordination was done by the GPD through the state governments. The defined project objectives were ensured by the GPD through appraisal of each project component submitted by the implementing agency. However, the involvement of aid agencies, with their associated mandatory procedures, also added to the complexities of decision-making, especially in the large STP projects. Of the original 261 sub-projects, 95 per cent are now complete and functioning satisfactorily.

River Water Quality Monitoring

Right from its inception in 1986, GAP started a very comprehensive water quality monitoring programme by obtaining the data from 27 monitoring stations. Most of these river water quality monitoring stations were already existed under other programs and only required strengthening.

Technical help was also received for a small part of this programme from the Overseas Development Agency of UK in the form of some automatic water quality monitoring stations, the associated modeling software, training and some hardware. Monitoring programme is being run on a permanent basis using the infrastructure of other agencies such as CPCB and Central Water Commission (CWC) to monitor the data from 16 stations. To evaluate the result of this programme, an independent study of the water quality has also been awarded to separate the universities for different regional stretches of the river.

Future

Apart from the visible improvement in the water quality, awareness generated by the project is an indicator of its success. It has resulted in the expansion of the programme over the entire Ganga basin to cover the other polluted tributaries. The GAP has further evolved to cover all the polluted stretches of the major national rivers including a few lakes. Considering the huge costs involved the central and state governments have agreed in principle to each share half of the costs of the projects under the "National Rivers Action Plan".

Conclusions

The GAP is a successful example of timely action due to environmental awareness at the governmental level. Even more than this, it exhibits the achievement potential which is attainable by "political will". It is a model which is constantly being upgraded and improved in other river pollution prevention projects. This may be due to higher water consumption, less nutritious dietary habits, higher grit loads, fewer sewer connections, insufficient flows and stagnation leading to bio-degradation of volatile fractions in the pipes themselves.

The most important lesson learned was the need for control of pathogenic contamination in treated effluent. This could not be tackled before because of the lack of safe and suitable technology but is now being attempted through research and by developing a suitable indigenous technology, which should not impart traces of any harmful residues in the treated effluent detrimental to the aquatic life.

4.2.2 Sustainable Life Studies

The definition of "sustainability" is the study of how natural systems function, remain diverse and produce everything it needs for ecology to remain in balance. It also acknowledges that human civilization takes resources to sustain modern way of life. There are countless examples throughout human history where a civilization has damaged its own environment and seriously affected its own survival chances. Sustainability takes into account how we might live in harmony with natural world around us, protecting it from damage and destruction.

We now live in a modern, consumerist and largely urban existence throughout the developed world and we consume a lot of natural resources every day. In our urban centres, we consume more power than those who live in rural settings and urban centres uses more power than average, keeping our streets and civic buildings lit, to power our appliances, our heating and other public and household power requirements. It is estimated that we use about 40% more resources every year than we can put back and that needs to change. Sustainability and sustainable development focuses on balancing that fine line between competing needs i.e., our need to move forward technologically and economically and the needs to protect the environments in which we and others live. Sustainability is not just about the environment, it is also about our health as a society in ensuring that no people or areas of life suffer as a result of environmental legislation and it is also about examining the longer term effects of the actions that humanity takes and asking questions about how it may be improved.

The Three Pillars of Sustainability

In 2005, the World Summit on Social Development identified three core areas that contribute to philosophy and social science of sustainable development. These "pillars" in many national standards and certification schemes, form the backbone of tackling the core areas that the world now faces. The Brundtland Commission described it as "development that meets the needs of the present without compromising the ability of future generations to meet their own needs". We must consider the future then, in making our decisions about the present.

Economic Development

It is the issue that proves the most problematic as most people disagree on political ideology what is and is not economically sound and how it will affect businesses and by extension, jobs and employability. It is also about providing incentives for businesses and other organizations to adhere to sustainability guidelines beyond their normal legislative requirements. Also, to encourage and foster incentives for the average person to do their bit where and when they can, one person can rarely achieve much, but taken as a group, effects in some areas are cumulative. The supply and demand market is consumerist in nature and modern life requires a lot of resources every single day, for the sake of environment, getting what we consume under control is the paramount issue. Economic development is about giving people what they want without compromising quality of life, especially in the developing world and reducing financial burden and "red tape" of doing the right thing.

Social Development

There are many facets to this pillar. Most importantly is awareness of and legislation protection of the health of people from pollution and other harmful activities of business and other organizations. In North America, Europe and the rest of the developed world, there are strong checks and programmers of legislation in place to ensure that the people's health and wellness is strongly protected. It is also about maintaining the access to basic resources without compromising the quality of life. The biggest hot topic for many people right now is sustainable housing and how we can better build the homes we live in from sustainable material. The final element is education i.e., encouraging people to participate in environmental sustainability and teaching them about the effects of environmental protection as well as warning of the dangers if we cannot achieve our goals.

Environmental Protection

We all know what we need to do to protect the environment, whether that is recycling, reducing our power consumption by switching the electronic devices off rather than using standby, by walking short journeys instead of taking the bus. Businesses are regulated to prevent pollution and to keep their own carbon emissions low. There are incentives to installing renewable power sources in our homes and businesses. Environmental protection is the third pillar and to many, the primary concern of the future of humanity. It defines how we should study and protect ecosystems, air quality, integrity and sustainability of our resources and focusing on the elements that place stress on the environment. It also concerns how technology will drive our greener future; the EPA recognized that developing technology is key to this sustainability and protecting the environment of the future from potential damage that technological advances could potentially bring.

4.2.3 Impact of Fire Crackers on Men and his Well Being

There are some adverse effects of crackers on health:

Air Pollution

Air pollution takes the first position when it comes to the adverse effects of crackers on health. Air pollution goes up by 50% during Diwali in India. Air pollution leads to respiratory illnesses such as Asthma, Bronchitis and other Lung problems. Hence, it is advised to avoid crackers to avert the hazards of air pollution on our health.

Smog

The pollution thus caused due to crackers also leads to Smog (Smoke + Fog). Smog is not only injurious to health but also causes problems during night driving as it obstructs our view. Smog occurs if there is no wind or rain within 2-3 days from such special occasion. Smog also leads to health problems such as skin burning and Lung Cancers. Hence, Smog is one of the most adverse effects of crackers on health.

Global Warming

The Oxides and Dioxides of Sulphur and Nitrogen are released during the burning of crackers. These are very much harmful to human health and also to the environment around us. These gases are also known as the "Green House Gases" which not only aid in the rising temperatures across the globe but also in the depletion of Earth's protective shield "The Ozone Layer" which might be hazardous to the living beings around us as they will be directly exposed to Sun's harmful radiations. Hence Crackers cause Global warming and Global warming has adverse effects on our health.

Noise Pollution

Noise pollution is as dangerous as the Air pollution. Noise pollution is also one of the leading adverse effects of crackers on health. It not only affects the Human Beings but also the animals around us such as cats and dogs as they have Ears that are more sensitive than the Humans. Noise pollution caused by crackers might also lead to temporary or permanent deafness.

Injuries and Wounds

Crackers are not only injurious to health but can also cause injuries to the individuals, especially to the children. If crackers are not fired properly they might cause wounds on the parts of body which is exposed to them. In many cases, children have lost their lives or have lost their eyesight by not lighting the crackers properly. Hence, it is one of the most adverse effects of crackers.

Garbage Disposals

After Diwali the main problem that comes to our sight is the Garbage lying on the roads. This garbage is chemical hazardous garbage that is very difficult to dispose of as it might affect the people's health that lives near these disposal areas of the cities. Another adverse Effect of crackers on health is that many firecracker factories and industries in India do not impose security measures and make people work in factories without any safety measures. These chemicals get exposed to their skin, eyes, lungs and increase toxic levels in the body. Many studies have shown that continuous exposure to such hazardous chemicals can lead to impairment or even impotency leading to birth of unhealthy babies.

Additional Health Hazards

Crackers are not only injurious to elderly citizens and children but also to the pregnant women. There have been many cases across the globe where the pollution has led to miscarriage in pregnant women. Hence pregnant women are advised not to leave their homes when the bursting of crackers is at its peak. Miscarriage is one of the most adverse effects of crackers on health. Other than this, many children are exploited and made to work in firecrackers factories. These children get exposed to harmful chemicals which hinders their growth or even increases toxic levels in their bodies.

Fire and Loss of Lives

During Diwali there are always headlines in the National Newspapers that some house/workshop has been a prey to fire due to crackers and many people lost their lives. Hence fire crackers not only injure us but can also kill us. Hence Fire is one of the most adverse effects of crackers on health and our lives too.

Therefore everyone must avoid crackers on Diwali or any other special occasion as its just equivalent to burning money. One can also celebrate it by having a get together with relatives and friends and sharing light and warm moments with them. Avoiding crackers will also protect our loved ones and will bring the family together during such occasions.

4.3   Solid Waste Management

The rapid population growth and urbanization in developing countries has led to people generating enormous quantities of solid waste and consequent environmental degradation. The waste is normally disposed in open dumps creating nuisance and an environmental degradation. Solid wastes cause a major risk to public health and the environment. Management of solid wastes is important in order to minimize the adverse effects posed by their indiscriminate disposal.

Types of Solid Wastes

Depending on the nature of origin, solid wastes are classified into:

- Urban or Municipal wastes.
- Industrial wastes.
- Hazardous wastes.

Sources of Urban Wastes

Urban wastes include the following wastes.

Domestic Wastes

It contains a variety of materials thrown out from homes.

Example: Food waste, Cloth, Waste paper, Glass bottles, Polythene bags, Waste metals, etc.

Commercial Wastes

It includes the wastes that comes out from shops, markets, hotels, offices, institutions, etc.,

Example: Waste paper, packaging material, cans, bottle, polythene bags, etc.

Construction Wastes

It includes wastes of construction materials.

Example: Wood, Concrete, Debris, etc.

Biomedical Wastes

It includes mostly waste organic materials.

Example: Anatomical wastes, Infectious wastes, etc.

Classification of Urban Wastes

Biodegradable Wastes

Those wastes that can be degraded by the microorganisms are called biodegradable wastes.

Example: Food, vegetables, tea leaves, dry leaves, etc.

Non-biodegradable Wastes

An urban solid waste material that cannot be degraded by microorganisms are called nonbiodegradable wastes.

Example: Polythene bags, scrap materials, glass bottles, etc.

Sources of Industrial Wastes

The main source of industrial wastes is chemical industries, metal and mineral processing industries.

Example: Thermal power plants produces fly ash in large quantities, Nuclear plants generated radioactive wastes, Chemical Industries produces large quantities of hazardous and toxic materials and Other industries produce packing materials, scrap metals organic wastes, rubbish, acid, alkali, rubber, plastic, dyes, paper, glass, wood, oils, paints, etc.

Effect of Improper Solid Waste Management

Due to the improper disposal of municipal solid waste on the roads and immediate surroundings, biodegradable materials undergo decomposition producing foul smell and become a breeding ground for disease vectors.

Industrial solid wastes are source for the toxic metals and hazardous wastes that affect the soil characteristics and productivity of the soils when they are dumped on the soil.

Toxic substances may percolate into the ground and contaminate the groundwater.

Burning of industrial or domestic wastes produce furans, dioxins and polychlorinated biphenyls which are harmful to human beings.

Solid waste management involves waste generation, mode of collection, transportation, segregation of wastes and disposal techniques.

Control Measures of Urban and Industrial Wastes

An integrated waste management strategy includes three main components such as:

- Source reduction.

- Recycling.

- Disposal.

```
┌─────────────────────────────────────────┐
│         ┌───────────────────────┐        │
│         │ Solid Wate Generation │        │
│         └───────────┬───────────┘        │
│                     ↓                     │
│          ┌────────────────────┐          │
│          │ Collection of Waste│          │
│          └─────────┬──────────┘          │
│                    ↓                      │
│           ┌────────────────┐             │
│           │ Transportation │             │
│           └───────┬────────┘             │
│                   ↓                       │
│              ┌─────────┐                 │
│              │ Storage │                 │
│              └────┬────┘                 │
│                   ↓                       │
│            ┌──────────────┐              │
│            │ Segregation  │              │
│            │  of Wastes   │              │
│            └──────┬───────┘              │
│                   ↓                       │
│           ┌─────────────────┐            │
│           │ Disposal methos │            │
│           └──┬──────┬──────┬─┘           │
│          ↙        ↓         ↘             │
│                b) Incineration           │
│    a) Landfills            c) Composting │
└─────────────────────────────────────────┘
```

Control measures.

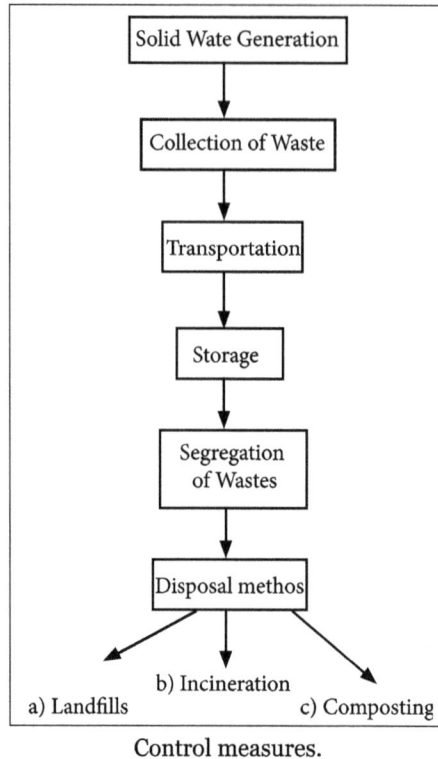

Source Reduction: Source reduction is a fundamental way to reduce waste. This can be done by minimizing the material when making a product, reuse of the products on site, designing products or packaging which helps to reduce their quantity. On an individual level we can reduce the use of unnecessary items while shopping, buy items which requires and has minimal packaging, avoid buying disposable items and also avoid using plastic carry bags.

Recycling: Recycling is reusing components of the waste that may have some economic value. Recycling has many visible benefits such as conservation of resources reduction in the energy used during manufacture and reducing the pollution levels. Some materials such as aluminum and steel can be recycled many times. Metal, paper, glass and plastics are recyclable.

Mining of the new aluminum is very expensive and hence recycled aluminum has a strong market and plays a significant role in the aluminum industry. Paper recycling can also help in preserving the forests as it needs about 17 trees to make one ton of paper. Crushed glass also known as cullet reduces the energy that is required to manufacture new glass by 50 percent. Cullet lowers the temperature requirement of glass-making process thus conserving energy and reducing air pollution.

However even if recycling is a viable alternative, it presents several problems. Problems associated with recycling are either technical or economical. Plastics are difficult to recycle because of different types of polymer resins used in their production. Since each type has its own chemical makeup different plastics cannot be recycled together. Thus separation of different plastics before recycling is necessary.

Similarly in recycled paper the fibers are weakened and are difficult to control the color of recycled product. The recycled paper is banned for use in food containers to prevent the possibility of contamination. It very often costs less to transport raw paper pulp than scrap paper. Collection, sorting and transport account for about 90 percent of the cost of paper recycling.

The processes of pulping, drinking and screening of wastepaper are generally more expensive than making paper from virgin wood or cellulose fibers. Very often thus recycled paper is more expensive than virgin paper. Although as technology improves the cost will come down.

Disposal: Disposal of solid waste can be done most commonly through a sanitary landfill or through incineration. A modern sanitary landfill is a depression in an impermeable soil layer that is lined with an impermeable membrane. The three key characteristics of a municipal sanitary landfill that distinguish it from an open dump are:

- Solid waste is placed in a suitably selected and prepared landfill site in a carefully prescribed manner.

- The waste material is spread out and compacted with appropriate heavy machinery.

Discarding Wastes

The following methods are adopted for discarding wastes:

- Landfill.

- Incineration.

- Composting.

Landfill: The solid wastes are placed in a sanitary landfill in which alternate layers of 80 cm covered with selected earth fill of 20 cm thickness. After 2-3 years solid waste volume shrinks by 25-30% and land is used for parks, roads and small buildings. This is the most common and cheapest cheapest method of waste disposal and is mostly employed in Indian cities.

Advantages:

- It is simple and economical.

- Segregation of wastes is not required.

- The landfilled areas can be reclaimed and used for other purposes.

- It converts low lying, marshy waste-land into useful areas.

- Natural resources are returned to soil and recycled.

Disadvantages:

- Large area is required.

- Land availability is away from the town, transportation costs are high.

- Leads to bad odour if landfill is not properly managed.

- Land filled areas will be sources of the mosquitoes and the flies requiring application of insecticides and pesticides at regular intervals.

- Causes fire hazard due to formation of methane in the wet weather.

Incineration: It is a hygienic way of disposing solid waste. It is suitable if waste contains more hazardous material and organic content. It is a thermal process and very effective for detoxification of all combustible pathogens. It is expensive when compared to composting or land filling.

In this method, municipal solid wastes are burnt in a furnace called as incinerator. Combustible substances such as rubbish organisms and garbage, dead combustible matter such as glass, porcelain and metals are separated before feeding to incinerators. The non-combustible materials can be left out for recycling and reuse. The leftover ashes and clinkers may account for about 10 to 20% which need further disposal by sanitary landfill or some other means.

The heat produced in the incinerator during burning of refuse is used in the form of the steam power for generation of electricity through turbines. The municipal solid waste is normally wet and has a high calorific value. Therefore, it has to be dried first before burning. Waste is dried in a preheater from where it is then taken to a large incinerating furnace known as the "destructor" which can incinerate about 100 to 150 tonnes per hour. The temperature normally maintained in the combustion chamber is about 700 C which may be increased to 1000 C when electricity is to be generated.

Advantages:

- Residue is only 20-25% of the original and can be used as clinker after treatment.

- Requires very little space.

- Cost of transportation is not high, then if the incinerator is located within the city limits.

- Safest from hygienic point of view.

- An incinerator plant of 3000 tonnes per day capacity can generate 3MW of power.

Disadvantages:

- Its capital and operating cost is high.

- Operation needs skilled personnel.

- Formation of smoke, dust and ashes needs further disposal and that may cause air pollution.

Composting: It is another popular method practiced in many cities in our country. In this method, bulk organic waste is converted into fertilizer by biological action.

Advantages:

- Manure added to soil increases water retention and ion-exchange capacity of soil.

- This method can be used to treat several industrial solid wastes.

- Manure can be sold thereby reducing cost of disposing waste.

- Recycling can be done.

Disadvantages:

- Non consumables have to be disposed separately.

- The technology has not caught-up with the farmers and hence does not have an assured market.

4.3.1 Consumerism and Waste Products

Consumerism refers, to the consumption of resources by the people. It is an organized movement of citizens and government. The special concentration is given to improve the rights and power of the buyers in relation to the sellers.

Consumerism is related to both increase in population as well as increase in our demand due to charge in lifestyle. In the modern society, our needs have a increased and so consumerism of resources has also increased.

Traditionally favorable rights of sellers:

- The right to introduce any product.

- The right to charge any price.

- The right to spend any amount to promote their product.

- The right to use incentives to promote their products.

Traditional buyer rights:

- The right to buy or not to buy.

- The right to expect a product to be safe.

- The right to expect the product to perform as claimed.

Important information's to be known by buyers:

- Ingredients of a product.

- Manufacturing date and expiry date.

- Whether the product has been manufactured against as established law of nature or involved in rights violation.

Objectives of consumerism:

- It improves the rights and power of the buyers.

- It involves making the manufacturer liable for the entire life cycle of a product.

- It forces the manufacturer to reuse and recycle the product after usage.

- The items which are very difficult to decompose like polymeric goods, computers, televisions etc., can be returned to manufacturer for reclaiming useful parts and disposing the rest.

- The reusable packing materials like bottles can be taken back to the manufacturer. It makes the products cheaper and avoids littering and pollution's.

- Active consumerism improves human health and happiness and also it saves resources.

Sources of Wastes

The sources of the waste materials are agriculture, mining, industrial and municipal wastes.

Examples for Waste Products

It includes glass, papers, garbage, plastics, soft drink canes, metals, food wastes, automobile wastes, dead animals, construction and factory wastes.

Effects of Wastes:

- The waste released from chemical industries and from explosives are dangerous to human life.

- The dumped wastes degrade soil and make unfit for irrigation.

- E-waste contains more than 1000 chemicals, which are toxic and cause environmental pollution. In computer, load is present in monitors, cadmium in chips and cathode ray tube, PVC in cables. All these cause cancer and other respiratory problems if inhaled for long periods.

- Plastics are difficult to recycle or incinerate safely because they are nonbiodegradable and their combustion produces several toxic gases.

Factors affecting consumerism and generation of wastes:

- People over Population: It occurs when there are more people than the available supply of food and water. Over population causes degradation of resources, poverty and premature death. This situation occurs in less developed countries. Thus in less developed countries per capita consumption of resources and waste generation are less.

- Consumption over Population: It occurs when there are less people than the available resources. Due to luxurious life-style per capita consumption of resources is very high. If the consumption is more, the generation of waste is also more and greater is the degradation of environment.

4.4   Bio-medical Waste

Bio-medical waste means "any solid and/or liquid waste including its container and any intermediate product, which is generated during the diagnosis, treatment or immunization of the human beings or animals.

The bio-medical waste poses hazard due to two principal reasons the first is infectivity and other toxicity.

The bio-medical waste consists of:

- The human anatomical waste like tissues organs and body parts.

- Animal wastes generated during the research from veterinary hospitals.

- Microbiology and biotechnology wastes.

- Waste sharps like hypodermic needles, syringes, scalpels and broken glass.

- Discarded medicines and cytotoxic drugs.

- Soiled waste such as dressing, bandages, plaster casts, material contaminated with blood, tubes and catheters.

- Liquid waste from any of the infected areas.

- The incineration ash and other chemical wastes.

Management and Handling Rules

At present, with the advancement of medical sciences most of the hospitals/nursing homes are well equipped with latest instruments for diagnosis and treatment of various diseases. One of the most important aspects associated with hospitals is the safe management of wastes, generated from these establishments, which contains human anatomical wastes blood, body fluid, disposable syringe, used bandages, surgical gloves, blood bags intravenous tubes etc.

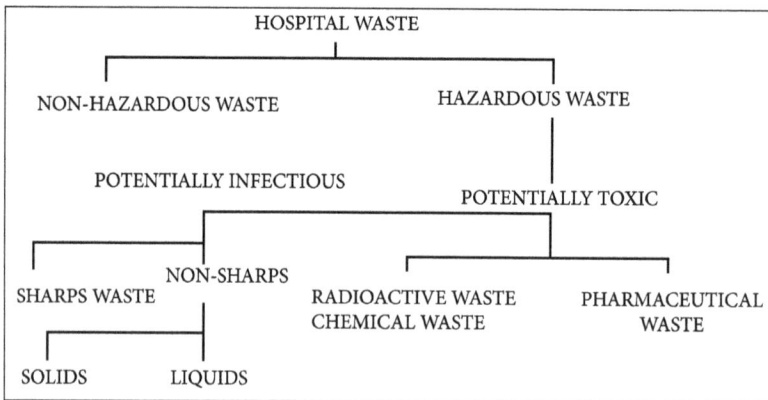

```
                              HOSPITAL WASTE

        NON-HAZARDOUS WASTE            HAZARDOUS WASTE

            POTENTIALLY INFECTIOUS
                                    POTENTIALLY TOXIC

                    NON-SHARPS
    SHARPS WASTE               RADIOACTIVE WASTE      PHARMACEUTICAL
                               CHEMICAL WASTE             WASTE

    SOLIDS      LIQUIDS
```

- The bio-medical waste generated from various sources has become a problem and much attention is being given worldwide to find out solution of this problem. The main concern lies with hospital waste generated from large hospitals/ nursing homes as it may pose deleterious effects due to its hazardous nature.

- The bio-medical wastes, if not handled in a proper way, is a potent source of diseases, like AIDS, Tuberculosis, Hepatitis and other bacterial diseases causing serious threats to human health. Owing to the potential threats this waste needs prime attention for its safe and proper disposal.

Sources of Bio-medical Waste

While urban solid waste has attracted attention of town planners, environmental activists and civic administrators, there is yet lack of concern for bio-medical waste generated primarily from health care establishments, including hospitals, nursing homes, veterinary hospitals, clinics and general practitioners, dispensaries, blood banks, animal houses, pathological laboratories and research institutes. The other sources of bio-medical waste are the following:

Households: The domestic sector generates biomedical waste to a small extent which is less than about 0.5% of the total waste generated in a household. The types of bio-medical waste produced in a household are syringes, cotton swabs, discarded medicines, bandages, plaster, sanitary napkins, diapers etc. The households are not covered under the law for generating bio-medical waste.

Industries, Education Institutes, Research Centers, Blood Banks, Health care establishments (for humans and animals) and Clinical Laboratories.

All the above mentioned organizations generate Bio-Medical Waste in substantial quantities. This sector is covered under purview of Bio-Medical Waste (Management and Handling) Rules.

The type of waste generated from an animal house is usually animal tissues organs, body parts, carcasses, body fluids, blood etc., of experimental animals. Besides this highly hazardous microbiological and biotechnology wastes is also produced.

Segregation: Segregation refers to basic separation of different categories of waste generated at source and thereby reducing the risks as well as cost of handling and disposal. It is the most crucial step in bio-medical waste management. Effective segregation alone can ensure effective bio-medical waste management.

- It reduces the amount of waste needs special handling and treatment.

- Effective segregation process prevents the mixture of medical waste like sharps with the general municipal waste.

- It prevents illegal reuse of certain components of medical waste like used syringes, needles and other plastics.

- It provides an opportunity for recycling certain components of medical waste like plastics after proper and thorough disinfection.

- The recycled plastic material can be used for non-food grade applications.

- Of the general waste, the biodegradable waste can be composted within the hospital premises and can be used for gardening purposes.

- Recycling is a good environmental practice, which can also double as a revenue generating activity.

- It reduces the cost of treatment and disposal (80 percent of a hospital's waste is general waste, which does not require special treatment, provided it is not contaminated with other infectious waste).

4.4.1 Hazardous and E–waste Management

Like hazardous waste, the problem of e-waste has become an immediate and long term concern as its unregulated accumulation and recycling can lead to major environmental problems endangering human health. Information technology has revolutionized the way we live, work and communicate bringing countless benefits and wealth to all its users. The creation of innovative and new technologies and the globalization of the economy have made a whole range of products available and affordable to the people changing their lifestyles significantly. New electronic products have become an integral part of our daily lives providing us with more comfort, security, easy and faster acquisition and the exchange of information. But on the other hand, it has also led to unrestrained resource consumption and alarming waste generation. Both developed countries and developing countries like India face the problem of e-waste management. Rapid growth of technology, upgradation of technical innovations and a high rate of obsolescence in electronics industry have led to one of the fastest growing waste streams in the world which consist of end of life electrical and electronic equipment products. It comprises a whole range of electrical and electronic items such as the refrigerators, washing machines, computers and printers, televisions, mobiles, i-pods, etc., many of which contain toxic materials. Many of the trends in consumption and production processes are unsustainable and pose serious challenge to environment and human health. The optimal and efficient use of natural resources, minimization of waste, development of cleaner products and environmentally sustainable recycling and disposal of waste are some of the issues which need to be addressed by all concerned while ensuring the economic growth and enhancing the quality of life. The countries of the European Union (EU) and other developed countries to an extent have addressed the issue of e-waste by taking policy initiatives and by adopting scientific methods of recycling and disposal of such waste. The EU defines this new waste stream as 'Waste Electrical and Electronic Equipment' (WEEE). As per its directive, the main features of the WEEE include definition of 'EEE', its classification into 10 categories and its extent as per voltage rating of 1000 volts for alternating current and 1500 volts for direct current. The EEE has been further classified into 'components', 'sub-assemblies' and 'consumables'. Since there is no definition of the WEEE in the environmental regulations in India, it is simply called 'e-waste'. E-waste or electronic waste, therefore, broadly describes loosely discarded, surplus, obsolete, broken, electrical or electronic devices.

In most part of the world, underground water is not drinkable directly. Long ago, people simply used to draw up water from wells and drink it. But now, we have to use some sort of filter to purify the water and make it drinkable. It is just one of the many problems and hazards of E-waste. The electronic devices, dead cells and batteries we throw away with other garbage contain lead that easily mixes with underground water, making it unfit for direct consumption.

E-Waste (Electronic and Electric Wastes)

Electronic equipment's like computers, printers, mobile, phones, xerox machines, calculators, etc. After using these Instruments, they are thrown as waste.

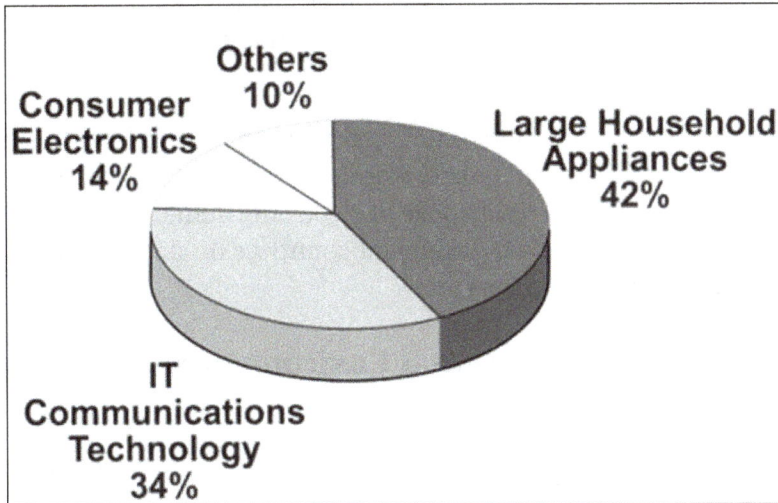

This word has caught up in the recent past only when someone studying the subject noted that our environment will be 3x more congested with e-waste by 2017. Even if it is not to be tripled, the e-waste is growing in volumes huge volumes. The reason why e-waste is increasing is that technology is growing fast and in an attempt to get better devices, we casually get rid of old electronics the best examples being that of smart phones.

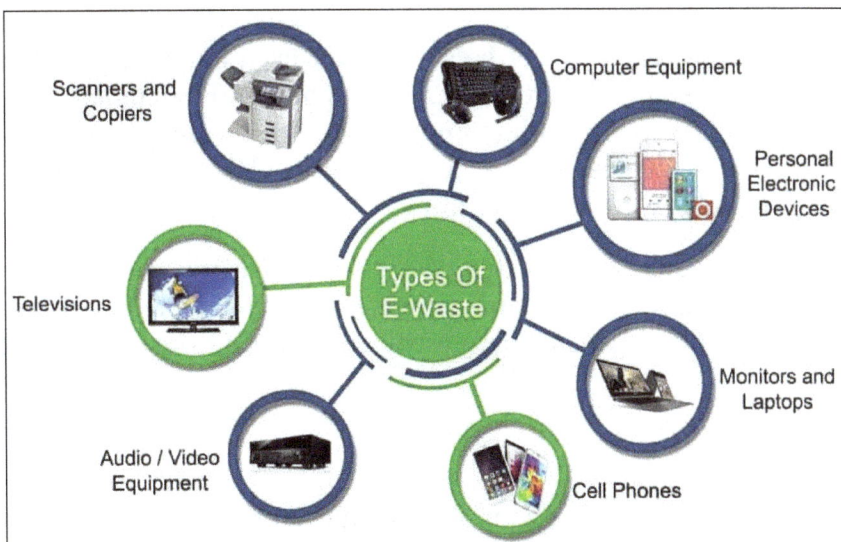

One may ask the relationship between old electronics and e-waste. E-waste is actually the old electronic goods that people simply give away to garbage trucks that are then dumped into landfill or similar sites. The electronics have a number of harmful elements that react with air and water to create problems of e-waste such as water, air and soil pollution as well as problems that affect human beings in the form of diseases.

Most of the cheaper batteries are lead based and easily react with water to seep and mix with underground water along with polluting the soil and air where it was disposed by garbage department.

Thus, everything that falls into electronics' category, that we intend to throw away, is e-waste (electronic waste). This includes computers, laptops, tablets, smart phones and so on. There are proper methods to dispose of electronic items. They should be handled differently, but unfortunately, even developed countries do not have strong policies to take care of such harmful, toxic garbage.

Effect of E-Waste on Humans and Environment

The solder present on the motherboard of computers and TV contain high levels of lead. Even the glass panels of the computer monitors and of course, the lead batteries contaminate air, water and soil. In addition, they distort the process of brain development, while posing danger to central nervous system and kidneys. This (lead poisoning) is among most dangerous hazards of e-waste.

Other than lead, motherboards also have high levels of Mercury. Improper disposal may create skin and respiratory disorders. Mercury poisoning also causes acute brain damages.

The cables and PVC panels as well as glass, when reacts with moisture and oxygen, creates hazardous soil that may not be suitable for even building a home as people breathing that air will suffer from reproduction and proper development of body parts, including brain. It also spoils the immune system. Stress, anxiety and other mental problems can arise out of breathing air polluted with glass, PVC and other forms of plastic remains found in electronic items.

The motherboard circuits can cause lung cancer when we breathe air polluted by the fumes released when the motherboard elements react and create Beryllium. It is also responsible for skin diseases, including warts and certain forms of dangerous allergies.

Treating E-Waste

There are no proper methods being implemented even in the first world to eliminate the problem of e-waste. The two methods for proper treatment of e-waste are recycling and refurbishing.

For recycling, there may be products that cannot be recycled completely. PVC layers, for example, stay as such for ages and cannot be recycled. It would be better if manufacturers use recyclable material so that the e-waste is converted into something that can be used again without harming the planet and its inhabitants. Thus, one of the major factors in treating e-waste is to compel manufacturers to use green elements.

If electronics are refurbished, they can be sold again at a lower price. Thus, both the society and environment will benefit. Instead of simply dumping our old TV into the garbage bin, we might want to think about calling the vendor and ask him where to present the item for refurbishing. If we cannot find, consider donating the item to some charity that can either use it as such or get it repaired and use it.

The bottom line when talking about proper disposal of e-waste is to convert them into less harmful items before disposing them off completely. There should be a sound policy on this subject and should be implemented without any irregularities for the benefit of the entire planet.

Chapter 5

Social Issues and the Environment

5.1   Urban Problems Related to Energy

Urban center use enormous quantities of energy. In the past, urban housing required relatively smaller amounts of energy than we use at present. Traditional housing in India had required very little temperature adjustments as the material used, such as wood and bricks, handled temperature changes better than the current concrete, glass and steel of ultramodern building.

Cities are the main centers of economic growth, trade, education, innovations and employment. Until recently a large majority of human population lived in rural areas and their economic activities cantered on agriculture, fishing, cattle, rearing, hunting or some cottage industry.

It was about two hundred years ago with the dawn of industrial are the cities showed rapid development. Now about 50% of world population lives in the urban areas and there is increasing movement of rural folk to cities in search of employment.

The urban growth is very fast that it is becoming difficult to accommodate all the industrial, commercial and residential facilities within a limited municipal boundary. As a result there is spreading of cities into the suburban or rural areas too, this phenomenon is referred as "urban sprawl".

In developing countries too, urban growth is very fast and in most of the cases, it is uncontrollable and unplanned growth. In contrast to the rural set up, Urban set up is densely populated, consumes a lot of energy and materials and generates a lot of waste.

Energy use is closely related to development in industry, communication, transport, commercial, household and agricultural activities. Energy requirement of urban population is much higher than that of rural ones. This is because urban people have a higher standard of life and their lifestyle demands more energy inputs in every sphere of life.

In urban areas, the need of energy is increasing by leaps and bounds. Furthermore, countries use energy in an uneven manner in the world. In developed countries, the amount of energy used is much more compared to developing countries.

5.1.1 Water Conservation

The following strategies can be adopted for conservation of water:

1. Reducing evaporation losses

Evaporation of water in humid region can be reduced by placing horizontal barriers of asphalt below the soil surface, which increases the water availability and crop yield.

2. Reducing irrigation losses

The water losses during irrigation can be reduced by the following methods.

- Sprinkling irrigation and drip irrigation conserves water by 30-40%.

- Growing hybrid crop varieties, which require less water, also conserve water.

- Irrigation in early morning or later evening reduces evaporation losses.

3. Reuse of water

- Treated wastes can also be used for fertile irrigation.

- Grey water from washing's, bathrooms, etc., may be used for wasting cars, watering gardens.

4. Preventing wastage of water

Wastage of water can be prevented by:

- Closing the taps when not in use.

- Repairing any leakage from pipes.

- Using small capacity of taps.

5. Decreasing run-off losses

Run off on most of the soils can be reduced by allowing most of the water to infiltrate into the soil. This can be done by using contour cultivation or terrace farming.

6. Avoid discharge of sewage

The discharge of sewage into natural water resources should be prevented as much as possible.

5.1.2 Rain Water Harvesting

Rainwater harvesting is the accumulation and deposition of rainwater for reuse on site, rather than allowing it to run off. Its uses include water for garden, water for livestock, indoor heating, water for irrigation, water for domestic use with proper treatment and for houses etc.

In many places the water collected is just redirected to a deep pit with percolation. The harvested water can be used as drinking water as well as for storage and other purpose like irrigation.

Objectives of Rain Water Harvesting:

- To meet the increasing demands.

- Raise the water table by recharging the ground water.

- Reduce ground water contamination.

- Reduce the surface run off loss & soil erosion.

- Increase in hydro static pressure.

- Minimize water crisis & water conflicts.

The following figure shows the working system:

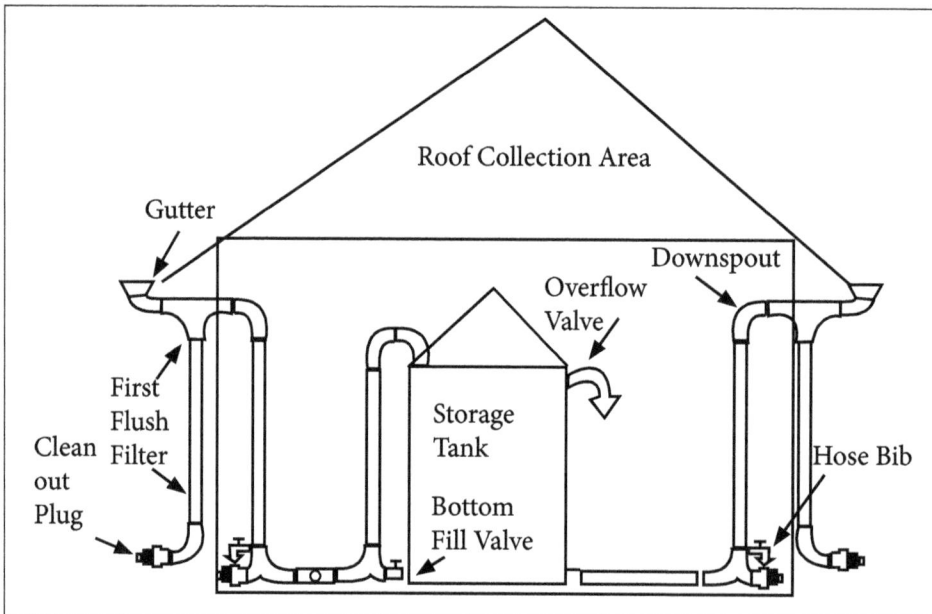

Rain water harvesting.

The advantages of rain water harvesting are:

- Reduction in the use of current for pumping water.

- Migrating the effects of droughts and achieving drought proofing.

- Increasing the availability of water from well.

- Rise in ground water level.

- Minimizing the soil erosion and flood hazards.

- Upgrading the social and environmental status.

- Future generation assured of water.

Waste Shed Management

The source of the waste material is agriculture, mining, industrial and municipal wastes.

Objectives:

- To minimize of risk of floods.

- For improving the economy.

- For developmental activities.

- To generate huge employment opportunities.

- To promote forestry.

- To protect soil from erosion.

Water shed management Techniques:

- Trenches (Pits).

- Dam.

- Farm pond.

- Underground barriers.

Maintenance of Watershed:

- Water harvesting.

- Afforestation.

- Reducing soil erosion.

- Scientific mining & Quarrying.

- Public participation.

- Minimizing livestock population.

5.1.3 Resettlement and Rehabilitation of People, Problems and Concerns

Rehabilitation of People

People are forced to evacuate from their land due to both natural and man-made disasters. Natural disasters like earthquakes, cyclones, tsunami etc.

Render thousands of people homeless and sometime even force them to move and resettle in other areas. Similarly, developmental projects like construction of dams, roads, flyovers and canals displace people from their home. Thus, resettlement refers to the process of settling again in a new area. Rehabilitation means restoration to the former state.

Resettlement Issues

As per the World Bank estimates, nearly 10 lakhs people are displaced worldwide for a variety of reasons.

Little or no Support: Displacement mainly hits tribal and rural people who usually do not figure in the priority list of political authorities or parties.

Meager Compensation: The compensation for the land lost is often not paid; it is delayed or even if paid, is too small for both in monetary terms and social changes forced on them by these big developmental projects.

Loss of Livelihood: Displacement is not a simple incident in the lives of the displaced people. They should leave their ancestral land and forests on which they are depended for their livelihood. Many of them do not possess any skills to take up another activity or pick up any other occupation. Usually, the new land that is offered to them is of poor quality and refugees are unable to make a living.

Lack of Facilities: When people are resettled in a new area, basic infrastructure and amenities are not provided in that area. Very often, temporary camps become permanent settlements. It is also an important problem of displacement or resettlement that people have to face.

Increase in Stress: Resettlement disrupts the entire life of people. They are unable to bear the shocks of purposelessness and emptiness created in their life. The payment of compensation to the head of the family often lead to bitter quarrels over sharing of compensation amount within the family, leading to stress and even withering of family life. Moreover, the land ownership has a certain prestige attached to it that cannot be compensated for, even after providing the new land. With the loss of property and prestige, the marriages of young people also become difficult as people from the outside villages are not willing to marry their daughters to the refugees.

Increase in Health Problems: Lack of nutrition due to the loss of agriculture and forest based livelihood, lead to the general decline in the health of the people. People are used to traditional home remedies. But the herbal remedies and plants gets submerged due to the developmental projects.

Secondary Displacement: The occupational groups which are residing outside the submergence area but depending on the area for livelihood also experience unemployment. Laborers, village artisans, petty traders etc., lose their living.

Loss of Identity: Tribal life is based on the community. The tribal are very simple people who have a lifestyle of their own. Displacement has a negative impact on their livelihood, cultural and spiritual existence in the following ways.

- Inter community marriages, folk songs, cultural functions and dances do not take place among the displaced people. When they are resettled, it is usually individual based resettlement, which ignores communal character.

- Breakup of the families and communities are the important social issues of displacement. The women suffer the most as they are deprived even a little compensation.

- Resettlement increases the poverty of tribal due to the loss of land, livelihood, food insecurity, jobs, skills etc.

- Loss of the identity of individuals and loss of connection between the people and the environment is greatest loss in the process. The indigenous knowledge that they have regarding the herbal plants and wildlife are lost.

- The laws of land acquisition do not pay attention to the idea of communal ownership of property which increases stress within the family.

- The tribal people are not familiar with market trends, prices of commodities and policies. As such, they are exploited and get alienated in modem era.

Objectives of Rehabilitation: The following objectives of the rehabilitation should be kept in mind before people are given an alternative site for living:

- The tribal people should be allowed to live along the lives of their own patterns and others should avoid imposing anything on them.

- They should be provided means to develop their own traditional art and culture in possible way.

- The villagers should be given the option of shifting out with the others to enable them to live community based life.

- The people displaced should get an appropriate share in the fruits of the development.

- If resettlement is not possible in the neighbor area, priority should be given to development of the irrigation facilities and supply of basic inputs for agriculture, drinking water, wells, grazing ground for the schools for the children, cattle, primary health care units and other amenities.

- Removal of poverty should be one of the objectives of rehabilitation.

- The elderly people of village should be involved in the decision making.

- The displaced people should be given employment opportunities.

- Resettlement should be in the neighborhood of their own environment.

- Villagers should be taken into confidence at every stage of implementation of the displacement and they should be educated, through public meetings, discussion about the legalities of the Land Acquisition act and other rehabilitation provisions.

5.2   Environmental Ethics: Issues and Possible Solutions

Environmental ethics refers to the issues, principles and guidelines relating to human interactions with their environment. It also means that effort must be taken to protect an environment and to maintain its stability from the hazardous chemical pollutants.

Functions of environment:

- It is the life supporting medium for all organisms.

- It provides food, air, water and other important natural resources to the human beings.

- It disintegrates all the waste materials discharges by the modern society.

- It moderates the climatic conditions of the soil.

- A healthy economy depends on a healthy environment.

Environmental Problems:

- Deforestation activities.

- Population growth and urbanization.

- Pollution due to discharge of effluent and smoke discharge from the industries.

- Water scarcity.

- Land degradation and degradation of soil fertility.

Solutions to Environmental Problems

The environmental can be protected due to the following activities:

- Reduce the waste of matter and energy resources.

- Recycle and reuse as many of our waste products and resources as possible.

- Over exploitation of natural resources must be reduced.

- Soil degradation must be minimized.

- Sustainable development is essential on, conservation of resources, harvesting of non-conventional energy and waste management.

- Biodiversity of the earth must be protected.

- Reduce population and increase the economic growth of our country.

5.2.1 Environmental Protection Act and Air Act

Environmental Protection Act

This is a general legislation law in order to rectify the gaps and laps in the above Acts. This Acts empowers the central government to fix the standards for quality of air water, soil and noise and to formulate procedures and safe guards for handling of hazard substance.

Objective of environmental act:

- To protect and improve the environment.

- To prevent hazard to all living creatures and property.

- To maintain harmonious relationship between humans and their environment.

Important features of environment act:

- The Act further empowers the Government to lay down procedures and safe guards for the prevention of accidents which cause pollution and remedial measures if an accident occurs.

- The Government has the authority to close or prohibit or regulate any industry or its operation, if the violation for the provisions of the Act occurs.

- If the violation continuous, an additional fine of Rupees five thousands per day may be improved for the entire period of violation of rules.

- The Act fixes the liability of the offense punishable under Act on the person who is directly in change.

AIR (Prevention and Control of Pollution) ACT, 1981

This Act was enacted in the conference held at Stockholm in 1972. It deals with the problems relating to air pollution. It envisages the establishment of central and state control boards endowed with absolute powers to monitor air Quality.

Objectives of AIR act:

- To prevent, control and abatement of air pollution.

- To maintain the Quality of air.

- To establish a board for the prevention and control of air pollution.

Important features of AIR Act:

- The central board may lay down the standards.

- The central board co-ordinates and settle disputes between state boards, in addition to providing technical assistance.

- The state boards are empowered to lay down the standards for emissions of air pollutants from industrial units or automobiles.

- The state boards are to collect and disseminate information related to air pollution.

- The state boards are to examine the manufacturing orders.

- The state board can adhesive the state government to declare certain heavily polluted areas as pollution control areas.

- The directions of the central board are mandatory on state boards.

- The operation of an industrial unit is prohibited in heavily polluted areas.

- Violation of law is punishable with imprisonment for a term which may extend to 3 months.

5.2.2 Water (Prevention and Control of Pollution) Act

The Water Act was enacted by Parliament Act, 1974 for the prevention and control of water pollution and maintaining or restoring of wholesomeness of water.

The relevant provisions of this act are given below:

- Under section 21: Officials of pollution control board can take samples of water effluent from any industry, stream or well or sewage sample for the purpose of analysis.

- Under section 23: Officials of the state boards can enter any premises for the purpose of examining any record, plant, register or any of the functions of the Board entrusted.

- Under section 24: No person shall discharge any poisonous, noxious or any polluting matter into any stream or well or sewer or on land.

- Under section 25: No person without the previous consent to:

  ○ Establish or take any step to establish any industry, operation or process or any treatment and disposal system for any extension or addition there to, which is likely to discharge sewage or trade effluent into a stream or well or sewer or on land.

  ○ Bring into use any new or altered outlet for the discharge of sewage.

  ○ Begin to make any new discharge of sewage.

Under this section, the state board may grant consent to industry after satisfying itself on the pollution control measures taken by a unit or refuse such consent for some reasons to be recorded in writing.

- Under section 27: The state board may grant consent to the industry after satisfying itself on the pollution control measures taken by the unit or refuse such consent for reasons to be recorded in writing.

- Under section 28: Any person aggrieved by the order made by the State Board may within 30 days from the date, on which the order is communicated, prefer an appeal to such authority.

- Under section 33: State Board can direct any person who is likely to cause or has cause the pollution of water in street or well to desist from taking such action as is likely to cause its pollution or to remove such matters as specified by Board through court.

- Under section 33A: Pollution control board can issue any directions to any person, officer or authority and such person shall be bound to comply with such directions. The directions include the power to direct:

  ○ The closure, prohibition of any industry.

  ○ Stoppage or regulations of supply of electricity, water or any other services.

- Under section 43: Whoever contravenes the provisions of Section 24 shall be punishable with imprisonment for a term which shall not be less than one year and six months but which may extend to six years with fine.

- Under section 45: If any who has been convicted of any offense is again found guilty of an offense involving a contravention of the same provision shall be on the second and on every subsequent conviction be punishable with imprisonment for a term which shall not be less than two years but which may extend to seven years with fine.

- Under section 45 A: Whoever contravenes any of the provisions of this act or fails to comply with any order or direction given under this act for which no penalty has been elsewhere provided in this act, shall be punishable with imprisonment which may extend to 3 months or with fine which may extend to 10,000 rupees or with both.

Drawbacks of pollution related acts:

- Power and authority has been given only to the central government with a little power to the state government. This hinders effective implementation of the act in the states.

- The penalties imposed by the above act are very small when comparing to the damage caused by big industries due to pollution.

- A person cannot directly file a petition in the court.

- Litigation, related to the environment is expensive, since it involves technical knowledge.

- For small industries, it is very expensive to install an individual custom made effluent treatment plant.

5.2.3 Wildlife Protection Act

This act was amended in 1983, 1986 and 1991.

This act is aimed to protect and preserve wildlife. Wild life refers to all animals and plants that are not domesticated. India has rich wildlife heritage. It has 350 species of mammals, 1200 species of birds and about 20,000 known species of insects. Some of them are listed as 'endangered species' in the wildlife (Protection) act.

Wildlife is an integral part of our ecology and plays an essential role in its functioning. Wildlife is declining due to human actions, the wildlife products skins, furs, feathers, ivory, etc., have decimated the populations of many species.

Wildlife populations are regularly monitored and management strategies formulated to protect them.

Important features:

- The Act covers the rights and non-rights of forest dwellers.

- It provides restricted grazing in sanctuaries but prohibits in national park.

- It also prohibits the collection of non-timer forest.

- The rights of forest dwellers recognized by the Forest Policy of 1988 are taken away by the Amended Wild life Act of 1991.

Drawbacks of wildlife protection act, 1972:

- Since this act has been enacted just as fallout of Stockholm conference held in 1972, it has not included any locally evolved conservation measures.

- The ownership certificates for some animals often serve as a tool for illegal trading.

- Jammu and Kashmir have their own wildlife acts, therefore, trading and hunting of many endangered species and prohibited in other states are allowed in Jammu and Kashmir.

5.2.4 Forest Conservation Act

This act provides conservation of forests and related aspects. This act also covers all type of forests including reserved forests, protected forests and any forested land.

This Act is enacted in 1980. It aims at to arrest deforestation.

Important features of Forest Act:

- The reserved forests shall not be diverted or de reserved without the prior permission of the central government.

- This land that has been notified or registered or forest land may not be caused for non-forest purposes.

- Any illegal non forest activity within a forest area can be immediately stopped under Act.

Important features of Amendment Act of 1988:

- Forest departments are forbidden to assign any forest land by way of lease or otherwise to any private person or non-government body for reafforestation.

- Clearance of any forest land of naturally grown trees for the purpose of reafforestation is forbidden.

- The diversion of forest land for non-forest uses is cognizable offense and anyone who violates the law is punishable.

Drawbacks of Forest (conservation) Act, 1980:

- This act only transfers the powers from state to centre to decide the conversion of reserve forests to non-forest areas.

- The power has been centralized at the top and local communities have been completely ignored from the decision making process regarding the nature of forest areas.

- Tribal people living in forests are totally dependent on forest resources. If they are stopped from an exploiting forest for their livelihood and they resort to the criminal activities like killing, smuggling, etc.

- This law is concentrated on protecting trees, birds and animals but not on protecting poor people.

- The forest dwelling tribal communities have a rich knowledge about forest resources, their conservation and importance. However, their role and contribution is not being acknowledged.

5.3   Issues involved in Enforcement of Environmental Legislation

Three issues that are especially important for environmental legislation are as follows:

Freedom of information:

Environmental management and planning is hindered if the public, NGOs or even official bodies are unable to get the information.

A number of laws have been enforced for safeguarding environmental quality. Although, these laws and acts could not be enacted successfully.

The polluter pays principle:

The polluter pays for the damaged caused by a development this principle also implies that a polluter pays for monitoring and policing. A problem with this type of approach is that fines may bankrupt small businesses, yet be low enough for the large company to write them off as an occasional overhead, that does little for pollution control.

The precautionary principle:

This principle has evolved to deal with the risks and uncertainties faced by the environmental management. The principle implies that prevention is worth a pound of cure, it does not prevent problems but may reduce their occurrence and helps ensure contingency plans are made.

5.3.1 Public Awareness

It is evident that the growing number of poor people, in developing countries due to the rapid population growth with economic constraint which contributes to the degradation of environment and the renewable to the degradation of environment and the renewable sources like forests, water and extinction of various species on which man depends.

For these, greater awareness is needed. Care is necessary to harness the natural resources, so that the quality of the environment does not deteriorate.

One of the reasons for this is improper implementation of the various environmental laws and standards. The most important reason is due to lack of awareness and understanding the implicate environmental degradation.

Chapter 6

Environmental Management

6.1 Environmental Impact Assessment

The Environmental Impact Assessment (EIA) is not carried out rigidly it is a process comprising a series of steps. These steps are outlined below and the techniques more commonly used in EIA are described in techniques.

The steps in the EIA process are given below:

- Screening.

- Scoping.

- Prediction and mitigation.

- Management and monitoring.

- Audit.

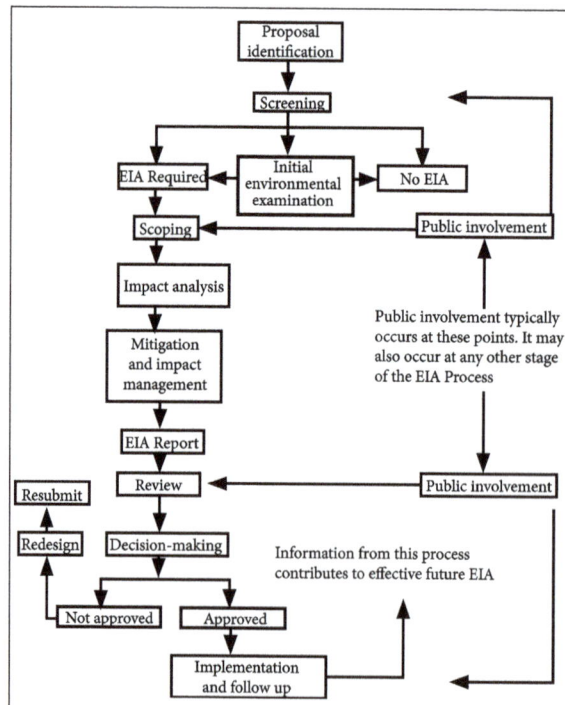

Generalized EIA Process Flow-Chart.

- Screening often results in a categorization of the project and from a decision is made on whether or not a full EIA should be carried out.

- Scoping is the process of determining which are the most critical issues to study and will involve the community participation to some degree. That EIA can most strongly influence the outline proposal.

- Detailed prediction and the mitigation studies follow scoping and are carried out in parallel with the feasibility studies.

- The main output report is known as an Environmental Impact Statement and contains a detailed plan for managing and monitoring the environmental impacts both during and after implementation.

- An audit of EIA process is carried out some time after implementation. The audit serves as a useful feedback and learning function.

Resources

An EIA team for an irrigation and drainage study is likely to be composed of some or all of the following namely a team leader, a hydrologist, an irrigation or drainage engineer, a fisheries biologist or ecologist, an agronomist/pesticide expert, a soil conservation expert, a biological or environmental scientist an economist, a social scientist and a health scientist. The final structure of team will vary depending on the project. Specialists may also be required for fieldwork, laboratory testing, library research, data processing, surveys and modeling.

There will be a large number of people involved in EIA apart from the full-time team members. These people will be based in a wide range of organizations, such as the project proposing and the authorizing bodies, regulatory authorities and various interest groups. Such personnel would be located in various agencies and also in the private sector; a considerable number will need specific EIA training.

Screening

Screening is the process of deciding on whether an EIA is required. This may be determined by size. Alternatively it may be based on the site specific information. For example, the repair of a recently destroyed diversion structure is unlikely to require an EIA whilst a new headwork structure may. Guidelines for whether or not an EIA is required will be country specific depending on laws or norms in operation. The legislation often specifies the criteria for screening and full EIA. All the major donors screen projects presented for financing to decide whether an EIA is required.

The output from screening process is often a document called an Initial Environmental Examination or Evaluation (IEE). The main conclusion will be as a classification of the project according to its likely environmental sensitivity. This will determine whether an EIA is needed and if so to what detail.

Scoping

Scoping occurs early in the project cycle at the same time as outline planning and pre-feasibility studies. Scoping is the process of identifying the key environmental issues and is perhaps the most important step in an EIA.

Scoping is important for two reasons. So that problems can be pinpointed early thereby allowing mitigating design changes to be made before expensive detailed work is carried out. To ensure that the detailed prediction work is only carried out for important issues. It is not the purpose of an EIA to carry out exhaustive studies on all the environmental impacts for all projects. If key issues are identified and a full scale EIA considered necessary then scoping should include terms of reference for these further studies.

At this stage the option exists for canceling or drastically revising the project should major environmental problems be identified. Equally it may be the end of the EIA process should the impacts be found to be insignificant. Once this stage has passed, the opportunity for major changes to the project is restricted.

Prediction and Mitigation

Once the scoping exercise is complete and major impacts to be studied have been identified, prediction work can start. This stage forms central part of an EIA. Several major options have been proposed either at the scoping stage or before and each option may require separate prediction studies. The realistic and affordable mitigating measures cannot be proposed without estimating the scope of the impacts first, which should be in monetary terms wherever possible. It then becomes important to quantify the impact of the suggested improvements by further prediction work.

An important outcome of this stage will be recommendations for the mitigating measures. This would be contained in the Environmental Impact Statement. Formal and informal communication links need to be established with teams that carryout by feasibility studies so that work can take the proposals into account. Similarly, feasibility studies may indicate that some options are technically or economically unacceptable and therefore, environmental prediction work for these options will not be required.

Many mitigating measures do not define physical changes but require management or institutional changes or additional investment, such as for health services. Mitigating measures may also be procedural changes, for example, the introduction of or increase in, the irrigation service fees to promote efficiency and water conservation.

By the time prediction and mitigation are undertaken, the project preparation can be advanced and a decision will most likely have been made to proceed with the project.

A considerable expenditure may have already been made and budgets allocated for the implementation of the project. Major changes could be disruptive to project processing and only accepted if prediction shows that impacts will be considerably worse than the originally identified at the scoping stage. To avoid conflict it is important that the EIA process commences early in the project cycle.

This phase of an EIA will work with the technical specialists with particular emphasis on:

- Interpretation of predictions, with and without mitigating measures.

- Prediction methods

- Assessment of comparisons.

Checklists, networks diagrams, matrices, graphical comparisons are techniques developed that to help carry out an EIA and present the results of an EIA in a format useful for comparing options. The main quantifiable methods of comparing options are by applying weightings, to environmental impacts. Numerical values or weightings can be applied to different environmental impacts to define their relative importance. However, economic techniques can provide insight into comparative importance where different environmental impacts are to be compared, such as either losing more wetlands or resettling a greater number of people.

When comparing a range of the proposals or a variety of mitigation or enhancement activities, characteristics of the different impacts needs to be highlighted. The relative importance of impacts needs agreeing, usually following a method of reaching consensus but including economic considerations. The uncertainty in predicting the impact should be clearly noted.

Management and Monitoring

The part of the EIS covering monitoring and management is often referred to as the Environmental Action Plan or Environmental Management Plan. Management and monitoring section not only sets out the mitigation measures needed for environmental management, both in the short and long term, but also the institutional requirements for the implementation. The term 'institutional' is used in its broadest context to encompass relationships:

- Motivated by socio psychological groups and individuals.

- Between individuals and groups involved in economic transactions.

- Established by law between the individuals and government.

- Developed to articulate legal, financial and administrative links among public agencies.

The above list highlights the breadth of options available for environmental management, namely: changes in law, changes in prices and changes in the governmental institutions and changes in culture which may be influenced by education and information dissemination.

One of the more straightforward and effective changes is to setup a monitoring programme with clear definition as to which agencies are responsible for data collection, collation, interpretation and implementation of management measures.

The purpose of monitoring is to compare predicted and actual impacts, particularly if impacts are either very important or the scale of the impact cannot be very accurately predicted. The results of monitoring can be used to manage environment, particularly to highlight problems early so that action can be taken.

The range of the parameters requiring monitoring may be broad or narrow and will be dictated by the 'prediction and mitigation' stage of the EIA. The typical areas of concern where monitoring is weak are: water quality, both inflow and outflow, stress in sensitive ecosystems soil, fertility, particularly salinization problems, water related health hazards, equity of water distributions and groundwater levels.

The use of satellite imagery to monitor changes in land use and the 'health' of the land and sea is becoming more common and can prove a cost effective tool, particularly in areas with poor access. Remotely sensed data have advantage of not being constrained by political and administrative boundaries. They can be used as one particular overlay in a GIS. However, the authorization is needed for their use, which may be linked to national security issues and may thus be hampered by the reluctant governments.

Monitoring should not be seen as an open ended commitment to collect data. If the need for monitoring ceases, data collection should cease. Conversely, monitoring may reveal the need for more intensive study and the institutional infrastructure must be sufficiently flexible to adapt to changing demands. The information obtained from monitoring and management can be extremely useful for future EIAs, making them both more efficient and accurate.

The Environmental Management Plan needs to not only include clear recommendations for action and the procedures for their implementation but must also define a programme and costs. It must be quite clear exactly how management and mitigation methods are phased with project implementation and when costs will be incurred.

Mitigation and the management measures will not be adopted unless they can be shown to be practicable and good value for money. The plan should stipulate that if, during project implementation, major changes are introduced or if the project is aborted, the EIA procedures will be restarted to evaluate the effect of such actions.

6.1.1 Preparation of EMP and EIS

Environmental Management Plan (EMP)

An Environmental Management Plan is a detailed plan and schedule of measures necessary to minimize, mitigate, etc. any potential environmental impacts identified by the EIA. Once the EIA significant impacts have been identified, it is essential to prepare an Environmental Management Plan.

An EMP should consist of a set of mitigation, monitoring and the institutional measures to be taken during the implementation and operation of the proposed project to eliminate adverse environmental impacts.

The EMP should also include the actions needed to implement these measures, including the following features:

- Mitigation based on the environmental impacts reported in the EIA, the EMP must describe with technical details each mitigation measure.

- The EMP should then include monitoring objectives that specifies type of monitoring activities that linked with the mitigation measures. Specifically, the monitoring section of the EMP provides the following:

  ○ The Specific description and the technical details of the monitoring measures That Includes The Parameters To Be Measured, Methods To Be Used, Sampling Locations, Frequency Of Measurements, Detection Limits And Definition Of Thresholds That Will Signal The Need For Corrective Actions.

  ○ Monitoring And Reporting Procedures To Ensure The Early Detection Of Conditions, that are necessary to particular mitigation measure and to furnish the information on the progress and results of mitigation.

- The EMP should also provide a specific description of institutional arrangements i.e. who is responsible for carrying out mitigating and monitoring measures.

- Additionally, the EMP should include an estimate of the costs of the measure and activities are recommended.

- EMP must be operative throughout whole Project Cycle.

- It should consider compensatory measures if mitigation measures are not feasible or cost effective.

Environmental Impact Statement (EIS)

The final EIA report is referred to as an Environmental Impact Statement (EIS). Most national environmental laws have specified the content of EIS must have multilateral and bilateral financial institutions have also defined what should be contained in an EIS. Ideally, content of an EIS should have the following:

- Description of environment.

- Significant Environmental Impacts.

- Legal, Policy and Administrative Framework.

- Executive Summary.

- Socio-economic analysis of Project Impacts.

- Description of the Proposed Project in detail.

- Identification and Analysis of Alternatives.

- Monitoring Program.

- Knowledge gaps.

- List of References.

- Mitigation Action/Mitigation Management Plan.

- Environmental Management Plan.

- Public Involvement.

- Appendices including:

  ◦ The reference documents, photographs, unpublished data.

  ◦ Notes of Public Consultation sessions.

  ◦ Consulting team composition.

  ◦ Terms of Reference.

6.1.2 Environmental Audit

In order to capitalize on the experience and the knowledge gained, the last stage of an EIA is to carry out an Environmental Audit sometime after completion of the project or the implementation of a programme. It will therefore usually be done by a separate team of specialists to that working on bulk of the EIA. The audit should include an analysis of the technical, procedural and decision-making aspects of the EIA.

Technical aspects include:

- The accuracy of predictions.

- The adequacy of the baseline studies.

- The suitability of mitigation measures.

Procedural aspects include:

- The fairness of the public involvement measures.

- The efficiency of the procedure.

- The degree of coordination of roles and responsibilities.

Decision making aspects include:

- The implications for development.

- The utility of process for decision making.

The audit will determine whether recommendations and requirements made by the earlier EIA steps were incorporated successfully into project implementation.

There are many kinds of audit which can be conducted alone or not. The audit can be concentrated on organization, emission, compliance with standards and regulation, maintenance, security, material balance, training, outside contractors.

The International Chamber of Commerce presents the different steps of an EA as follows:

Preaudit Activities

It includes:

- Selection and scheduling of facility to audit.

- Selection of audit team.

- Contact with the facility and planning of the audit.

Site Activities

It includes:

- Understanding of internal controls.

- Assessment of an internal controls.

- Gathering of an audit evidence.

- Evaluation of an audit findings.

- Report of findings to facility.

Post Audit Activities

It includes:

- Production of a draft report.

- Production of a final report.

- Preparation and an implementation of an action plan.

- Monitoring of action plan.

PRE-AUDIT ACTIVITIES	ACTIVITIES AT SITE	POST-AUDIT ACTIVITIES

SELECT AND SCHEDULE FACILITY TO AUDIT
Based on:
    -Selection critein
    -Priorities assigned
	STEP 1 : UNDERSTAND INTERNAL CONTROLS
	-Review Background information
	-Opening meeting
	-Orientation tour of plant
	-Review audit plan
	-Confirm understanding of internal control
		ISSUE DRAFT REPORT
		-Corrected closing report
		-Determine distribution list
		-Distribute draft report
		-Allow time for correction

SELECT AUDIT TEAM MEMBERS -Confirm their availability -Make travel and lodging arrangements -Assign audit responsibilities	**STEP 2 : ASSESS INTERNAL CONTROLS** -Identify strength and weaknesses of internal controls -Adapt audit plan and resource allocation -Define testing and verification strategies	**ISSUE FINAL REPORT** -Corrected draft report -Highlight requirements for action plan -Determine action plan preparation dead line
CONTACT FACILITY AND PLAN AUDIT -Discuss audit programme -Obtain background information -Administer questionnaire -Define scope -Determine applicable requirements -Note priority topics -Modify or adapt protocols -Determine resource needs.	**STEP 3 : GATHER AUDIT EVIDENCE** -Apply testing and verification strategies -Collect data -Ensure protocol steps are completed -Review all findings and observations -Ensure that all findings are factual -Conduct further testing if required	**ACTION PLAN PREPARATION AND IMPLEMENTATION** -Based on audit findings in final report
	STEP 4 : EVALUATE AUDIT FINDINGS -Develop complete list of findings -Assemble working papers and documents -Integrate and summarize findings -Prepare report for closing meeting	
	STEP 5 : REPORT FINDINGS TO PLAN -Present findings at closing meeting -Discuss findings with plan personnel	**FOLLOW UP ON ACTION PLAN**

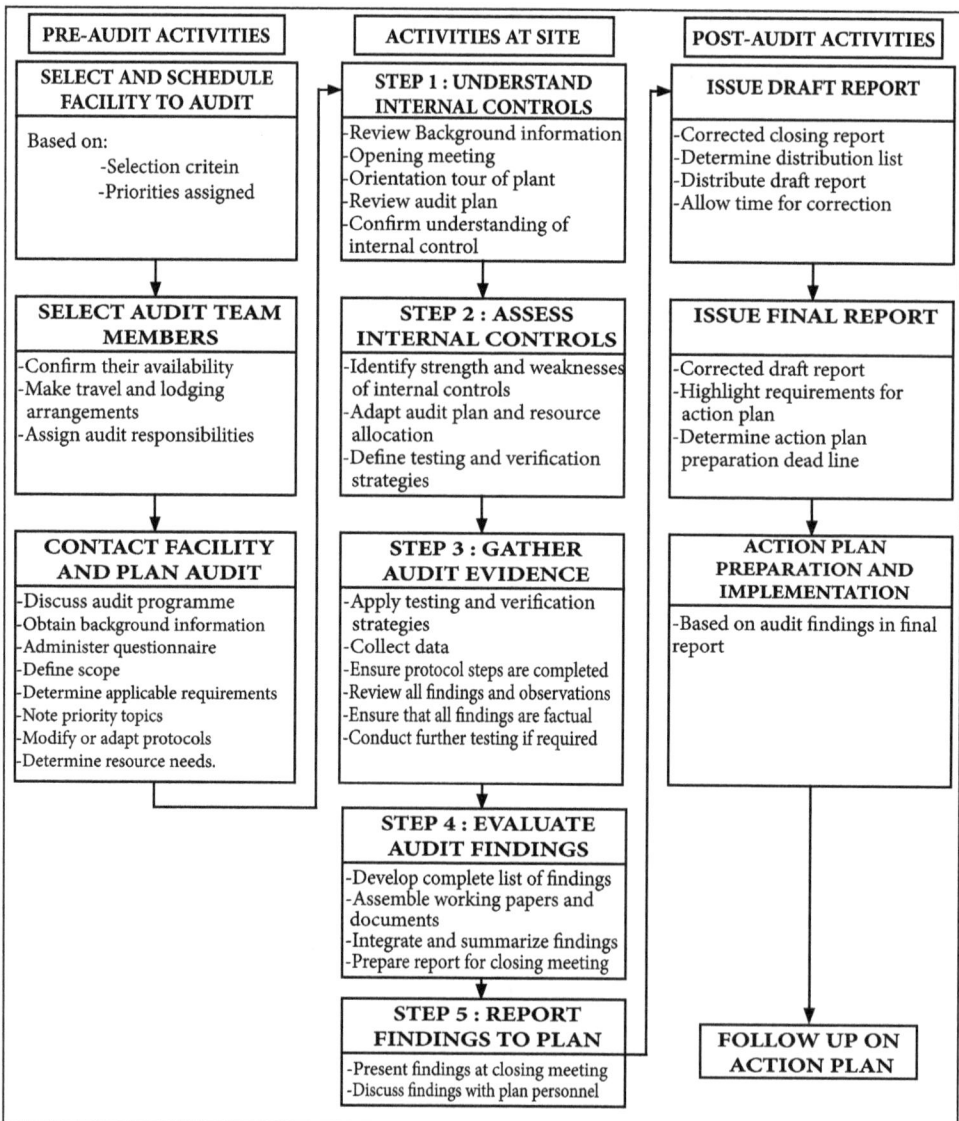

Basic steps of an Environmental Auditing.

6.2   Eco-tourism

Tourism involving travel to areas of natural or ecological interest, for the purpose of observing wildlife and learning about the environment. This is a conscientious form of tourism and tourism development. It encourages going back to the natural products in every aspect of life and it help to preserve nature. It is also the key to sustainable ecological development.

Principles of Eco-tourism

- Adopt low impact wildlife tourism that protects ecological integrity of forest.
- Highlight the biodiversity richness, their values and their ecological services to people.
- Build environmental and cultural awareness and respect.
- Facilitate that the sustainability of ecotourism enterprises and activities.
- Provide livelihood opportunities to local communities.
- Highlight the heritage value of India's wilderness and protected areas.

Issues Related to Environmental Studies

1. Climate Change

Global warming has been concerning scientists for many decades, but Al Gore legitimized the crisis with controversial film named "An Inconvenient Truth".

From the melting polar ice caps to catastrophic weather and threatened ecosystems, not only is climate change real, scientists agree that the humans are influencing climate change with our production of greenhouse gases.

2. Energy

Clean energy vs dirty energy. Renewable energy, Energy independence, Petroleum, Bio-fuels, Coal. ANWR and offshore drilling. Energy is the second only to climate change in significance.

Though no single energy source is going to be the solution, positive developments towards the clean future.

3. Waste

With the immediate looming problems of the climate change and energy, focus has shifted away from the landfill waste, but this is a serious problem. Water ways are choked with trash and modernized nations ship their undesirable leftovers to developing world.

The amount of waste the industrialized world generates is shocking. The water bottles are defining symbol of this critical issue. Luckily, people are becoming aware of the consequences of "fast consumption" and there are many simple changes we can make in our own life to help significantly reduce landfill waste.

4. Food

Biofuels have turned into a global controversy the idea that people may cause the starvation of millions in order to fuel their SUVs is disgusting and still that's not the whole picture. For example, eating hamburgers has as much as impact on the global food picture as the use of biofuels.

5. Water

Pure Water is in short supply. Our global reserves of drinkable water are a fraction of 1% and 1 in 5 humans does not have access to potable water. Many people do not realize that strife has already broken out in some of the stressed regions. There are many potential solutions, some promising, others challenging.

6. Consumption

This is directly tied to waste. It is well known that industrialized world simply consumes in a way that is not sustainable. And developing world is rapidly imitating the model. Sustainability in the most compelling sense is about long term solvency. The way we live now is against the future.

Reducing consumption and smart consumption are both necessary and there are many ways to go about doing this.

7. Land Management

From desertification to the polar ice melting to erosion and deforestation, the existing land management choices are not serving planet or its inhabitants very well.

1990s saw some headway with forest management but Bush administration's various initiatives have set back progress by decades.

8. Ecosystems and Endangered Species

Many species are now under threat, including indicator species and evolutionarily unique species. The consequences can have a global impact from the most unusual endangered animals to a complete list of indicator species for key.

6.2.1 Green Campus

It is one that carries out these functions according to a system-wide culture of environmental sustainability, balancing function and the design with existing and foreseen resources.

It is a place where environmentally responsible practice and education go hand in hand and where environmentally responsible tenets are borne out by example.

The green campus institution is a laboratory of self-scrutiny, experimentation and application. At its best, it is a model environmental community where the operational functions, business practices, academic programs and people are interlinked, providing educational and practical value to institution, the region and the world.

Green Campus concept offers institution the opportunity to take lead in rethinking its environmental culture and developing new paradigms for solving problems that are local, national and global in nature.

The greening initiatives are challenging and require determination and a long-term commitment on the part of entire campus community. These efforts, however, can yield significant paybacks. Such paybacks include:

- Environmental and economic sustainability: A system wide culture of sustainability helps preserve and enhance what we value as an institution today, as well as for the future.

- Reputation as a leader through example: Many colleges and universities fall far short of making their mark as environmental leaders, because they fail to practice on campus what they preach in the classroom. Although colleges and universities offer courses in the environmental management, engineering, laws and regulations and assessment, many have failed to comply with environmental requirements or to take part in pollution prevention activities. As a result, some institutions have been assessed substantial fines by U.S. Environmental Protection Agency (US EPA). Colleges and universities need to examine their own organizations and implement on their own campuses what they and the public expect industry to do.

- Economic benefits: A routine, curriculum-based, environmental audit program that reveals waste and inefficiency associated with campus activities, coupled with the identification of environmentally friendly alternatives, can yield significant cost savings for our institution. Without paying outside contractors, we can discover steps that institution can take to correct and improve environmental problems on campus. By acting on the recommendations resulting from these audits, colleges and universities can realize cost savings by reducing energy and water use, minimizing the campus waste stream, improving systems reliability and increasing the efficiency of heating/air conditioning systems. In addition, we will improve our environmental compliance and thereby minimize our potential for liability, fines and cleanup costs.

- "Real-life" work experience for students: The environmental audits and pollution prevention evaluations can be integrated into the curriculum, providing students with hands-on investigative and problem-solving experience that they can take with them when they enter workforce. This experience not only makes our students more marketable, it also provides them with the kinds of broad-thinking skills that allow them to succeed and thrive once they are employed.

- Improved quality of life on campus: A Green Campus is a cleaner, safer and healthier place to live and work.

6.2.2 Green Business and Green Politics

Green Business

Green businesses are socially and environmentally responsible: Green companies adopt principles and practices that protect people and the planet. They challenge themselves to bring the goals of social and economic justice, environmental sustainability, as well as community health and development, into all of their activities from production and supply chain management to employee relations and customer service.

Green businesses care for their workers: It ensures they do not use sweatshop or child labor. Everyone who works directly for them or their suppliers earns a living wage and works in healthy conditions. They create jobs that empower workers and honor their humanity. They also serve as models for the role businesses can play in the transformation of our society to one that is socially just and environmentally sustainable.

Green businesses protect their customers and clients: It ensures that they use the safest ingredients, to keep their customers and clients and their families healthy. They also provide green living alternatives to improve quality of life, with products and services that help in areas like affordable housing, sustainable agriculture, education, clean energy and efficiency, fair trade, healthy air, clean water and more. And they reduce, reuse and recycle, setting a good example.

Green businesses improve their communities: Along with ensuring their facilities are not polluting their local communities, many green businesses take steps to make the places that they call home better. Green businesses often spring up in marginalized communities' inner cities, rural and indigenous communities. Many are even started by the people in these communities who, in turn, bring respect and dignity to their employees and the wider neighborhood.

Green Politics

It is a relatively recent political movement that places a concern for nature and its myriad species at the top of its agenda. New social and political movements arise in response to crises that are perceived to be both long-term and systemic. The crisis out of which the broadly based Green movement has emerged is the environmental crisis, which is actually a series of interconnected crises caused by population growth, air and water pollution, the destruction of the tropical and temperate rain forests, the rapid extinction of entire species of plants and animals, the greenhouse effect, acid rain, ozone layer depletion and other now familiar instances of environmental degradation. Many are by-products of technological innovations, such as the internal combustion engine. But the causes of these environmental crises are not only technological but are also broadly cultural and political. They stem from beliefs and attitudes that place human beings above or apart from nature. Despite their differences, the major mainstream political perspectives liberalism, socialism and conservatism are alike in viewing nature as either a hostile force to be conquered or a resource base to be exploited for human purposes. All, in short, share an anthropocentric or human-centered, bias.

The modern environmental or Green movement counterpoises its own perspective. Many Greens prefer not to call their perspective a political ideology, but an environmental ethic. Earlier ecological thinkers, such as Aldo Leopold, spoke of a "land ethic". Others, such as Christopher stone, speak of an ethic with the earth itself at its center, while still others, such as Hans Jonas, speak of an emerging "planetary ethic". Despite differences of accent and emphasis, however, all are alike in several important respects. An environmental or Ecological ethics would include several key features. First, such an ethic would emphasize the web of interconnections and mutual dependence within which we and other species live. From this recognition of interconnectedness a second feature follows: a respect for all life, however humble humans may believe it to be, because the fate of our species is tied in with theirs. And since life requires certain conditions to sustain it, the third feature follows: we have an obligation to respect and care for the conditions that nurture and sustain life in its myriad forms. Since nature nourishes her creatures within a complex web of interconnected conditions, to damage one part of this life-sustaining web is to damage the others and to endanger the existence of the creatures that depend upon it.

Green thinkers hold that the enormous power that humans have over nature imposes on our species a special responsibility for restraining our reach and using our power wisely and well. Greens point out that the fate of the earth and all its creatures now depends, to an unprecedented degree, on human decisions and actions. For not only do we depend on nature, but nature depends on our care and restraint and forbearance. Humans have the nuclear means to destroy in mere minutes the earth's inhabitants and the ecosystems that sustain them. From this emerges a fourth feature of a "green" political perspective: Greens must oppose militarism and work for peace.

But the earth is in danger not only from global thermonuclear war but from the slower destruction of the natural environment. Such destruction is a consequence not only of large-scale policies but of small-scale, everyday acts. And each of us bears full responsibility for our actions and also, since we live in a democracy, some share of responsibility for cumulative effects and collective outcomes. Each of us has or can have a hand in making the laws under which we live. It is for this reason that Greens give equal emphasis to our collective and individual responsibility for protecting the environment that protects us. The fifth feature of the Green political perspective, then, is to emphasize the importance of informed and active democratic citizenship at the grass-roots level. Hence the Green adage, "Think globally and act locally."

On this much most Greens agree. But there are also a number of unresolved differences of approach, emphasis and political strategy. The internal ideological spectrum ranges from "light green" conservationists to "dark green" radicals and includes assorted anarchist beliefs, deep ecology, eco-feminism, social ecology, bioregionalism, New Age Gaia worship and other groupings, each differing in various ways form the others. Among these are differences regarding the basic beliefs underlying and motivating the green movement. Some New Age Greens envision an environmental ethic grounded in spiritual or religious values. We should look upon the earth as a benevolent and kindly deity the goddess Gaia (from the Greek word for "earth") to be worshiped in reverence and awe. In this way we can liberate ourselves from the restrictive rationalism that characterizes modern science. Other Greens disapprove of such a spiritual or religious orientation, contending that such beliefs are politically pernicious and inimical to the rational scientific thinking required to diagnose and solve environmental problems.

Other differences have to do with the political strategies and tactics to be employed by the environmental movement. Some say that Greens should take an active part in electoral politics, perhaps even following the lead of Greens in Germany and organizing a Green Party. Aware of the formidable obstacles facing minority third parties, most have favored other strategies, such as working within existing mainstream parties or hiring lobbyists to influence legislation. Still other Greens favor working outside of traditional interest group politics, believing the earth and its inhabitants hardly constitute a special interest. Others, such as social ecologists, tend to favor local, grass-roots campaigns which involve neighbors, friends and fellow citizens in efforts to protect the environment. Some social ecologists are anarchists who see the state and its pro-business and pro-growth policies as the problem, rather than the solution and seek its eventual replacement by a decentralized system of communes and cooperatives. Greens of the "bioregionalism" persuasion add that such social and political organization ought to be based on biological or natural, rather than artificial or political, boundaries and regions.

Although all Greens agree on the importance of informing and educating the public, they disagree as to how this might best be done. Some groups, such as Greenpeace, favor dramatic direct action calculated to make headlines and capture public attention. Even more militant groups have advocated monkey-wrenching as a morally justifiable means of publicizing and protesting practices destructive of the natural environment.

Such militant tactics are decried by moderate or mainstream groups, which tend to favor subtle, low key efforts to influence legislation and inform the public on environmental matters. The Sierra Club, for example, lobbies Congress and state legislatures to pass environmental legislation. It also publishes books and produces films and videos about a wide variety of environmental issues. Similar strategies are followed by other groups, such as the Environmental Defense Fund. Another group, The Nature Conservancy, solicits funds to buy land for nature preserves.

Differences over strategy and tactics are, however, differences about means and not necessarily about basic assumptions and ends. Despite their political differences, Greens are alike in assuming that all things are connected ecology is, after all, the study of interconnections and they agree that complex ecosystems and the myriad life-forms they sustain are valuable and worthy of protection by political and other means.

Permissions

All chapters in this book are published with permission under the Creative Commons Attribution Share Alike License or equivalent. Every chapter published in this book has been scrutinized by our experts. Their significance has been extensively debated. The topics covered herein carry significant information for a comprehensive understanding. They may even be implemented as practical applications or may be referred to as a beginning point for further studies.

We would like to thank the editorial team for lending their expertise to make the book truly unique. They have played a crucial role in the development of this book. Without their invaluable contributions this book wouldn't have been possible. They have made vital efforts to compile up to date information on the varied aspects of this subject to make this book a valuable addition to the collection of many professionals and students.

This book was conceptualized with the vision of imparting up-to-date and integrated information in this field. To ensure the same, a matchless editorial board was set up. Every individual on the board went through rigorous rounds of assessment to prove their worth. After which they invested a large part of their time researching and compiling the most relevant data for our readers.

The editorial board has been involved in producing this book since its inception. They have spent rigorous hours researching and exploring the diverse topics which have resulted in the successful publishing of this book. They have passed on their knowledge of decades through this book. To expedite this challenging task, the publisher supported the team at every step. A small team of assistant editors was also appointed to further simplify the editing procedure and attain best results for the readers.

Apart from the editorial board, the designing team has also invested a significant amount of their time in understanding the subject and creating the most relevant covers. They scrutinized every image to scout for the most suitable representation of the subject and create an appropriate cover for the book.

The publishing team has been an ardent support to the editorial, designing and production team. Their endless efforts to recruit the best for this project, has resulted in the accomplishment of this book. They are a veteran in the field of academics and their pool of knowledge is as vast as their experience in printing. Their expertise and guidance has proved useful at every step. Their uncompromising quality standards have made this book an exceptional effort. Their encouragement from time to time has been an inspiration for everyone.

The publisher and the editorial board hope that this book will prove to be a valuable piece of knowledge for students, practitioners and scholars across the globe.

Index